THE BETTER HALF

W/D

THE
BETTER
HALF

ON THE GENETIC
SUPERIORITY OF
WOMEN

SHARON MOALEM

ALLEN LANE
an imprint of
PENGUIN BOOKS

ALLEN LANE

UK | USA | Canada | Ireland | Australia
India | New Zealand | South Africa

Allen Lane is part of the Penguin Random House group of companies
whose addresses can be found at global.penguinrandomhouse.com.

Penguin
Random House
UK

First published in the United States of America by Farrar, Straus and Giroux 2020
First published in Great Britain by Allen Lane 2020
001

Printed and bound in Great Britain by Clays Ltd, Elcograf S.p.A.

A CIP catalogue record for this book is available from the British Library

ISBN: 978–0–241–39688–9

www.greenpenguin.co.uk

MIX
Paper from
responsible sources
FSC® C018179

Penguin Random House is committed to a
sustainable future for our business, our readers
and our planet. This book is made from Forest
Stewardship Council® certified paper.

For my better half

I have undertaken—as boldly as I can but not without shame—to treat a topic previously ignored but by no means far from the truth, namely, the nobility and preeminence of the female sex.

—HENRICUS CORNELIUS AGRIPPA, ANTWERP, APRIL 16, 1529

AUTHOR'S NOTE

Some of the names, case details, and identifying characteristics of people referenced in this book have been changed to protect the confidentiality of patients, colleagues, acquaintances, friends, and family. In some cases, scenarios and descriptions have been altered or combined to provide an added level of anonymity as well as to give clarity to an idea or diagnosis.

This book is intended as a reference volume only, not as a medical manual. It is not intended as a substitute for any treatment that may have been prescribed by your doctor. If you suspect that you have a medical problem, I urge you to seek competent medical help.

CONTENTS

THE BETTER HALF

INTRODUCTION

H ERE ARE SOME BASIC FACTS: Women live longer than
men. Women have stronger immune systems. Women are
less likely to suffer from a developmental disability, are more
likely to see the world in a wider variety of colors, and overall
are better at fighting cancer. Women are simply stronger than
men at every stage of life. But why?

I became fixated on this question one summer night as I
lay inside an ambulance speeding toward the hospital after a
serious car accident. Lying on the stretcher, hooked up to the

monitors, I found myself reflecting on two specific events in my past that had become vivid memories. One occurred when I was a doctor treating premature babies in a neonatal intensive care unit (NICU), and the other was ten years earlier, when I was focused on neurogenetics, working with people in their final years of life. The two events seemed connected in some way, but I couldn't quite put my finger on how.

Then, amid the chaotic activity in the back of that ambulance, the realization hit me. We all have those life events that make us question certain basic assumptions; the two things I thought about that summer night, and the crystallizing moment that followed, all link to the argument I'll be making in this book. And the thesis is this: women are genetically superior to men.

When I started doing research as a neurogeneticist (someone who specializes in the genetic mechanics of neurodegenerative disease), one of the unexpected challenges I experienced was recruiting sufficient numbers of healthy older adults to participate in the studies. Even with the perfect research questions and all the necessary financial support in place to test them, I'd often be stymied and have to delay because I couldn't find healthy older age- and sex-matched volunteers. The recruitment process could take years.

Unless, of course, you've got Sarah on your side. Sarah is in her late eighties and has two titanium hips, but with her walker she's pretty much unstoppable. Her weekly schedule consists of a watercolor course, swimming, and a cardio class, rounded out with a regular dancing soiree. If that isn't enough, Sarah takes part in almost daily events at different senior centers across the city. She's a member of a volunteer organization that visits hospitalized older adults who may not have any family or friends to spend time with them. She also happens to be my grandmother.

I'm often asked by family members if I'd be willing to speak to Sarah directly about slowing down. Everyone gets worried that she's simply too busy. My response to them is always the same: it's because she's so active and draws so much meaning from her daily activities that she's doing as well as she is. More important, if she stopped being social, I would quickly run out of older adult research volunteers.

My grandmother first started helping me recruit for my studies almost twenty years ago. She wasn't shy about giving advice either. "You'll never get one person wanting to help you with your study with that scary white lab coat and name tag on," she said. "If I were you, I'd try taking it off. And your nurse too—no lab coats. They scare us. It reminds me of my surgeries, and why would I want that? Without it, you can just look like a normal person. After all, you're asking people to give up a piece of themselves, and that's a big deal. You'll see—there are many people who want to help."

So I listened, and I ditched the lab coat. It worked. After I, dressed like a civilian, gave a presentation to prospective volunteers, we had more than the number of research participants we needed. The only problem was that even if everyone in the room agreed to participate, there would always be a glaring shortage of individuals in one specific demographic group. There just weren't enough men.

Elderly women on average outlive their male contemporaries by at least four to seven years. This longevity discrepancy becomes all the more striking as we start to approach the extreme end of the human life span. Over the age of eighty-five, women can outnumber men two to one. As for centenarians, women's survival advantage is even more exaggerated: out of one hundred centenarians alive today, *eighty* are women and only twenty are men.*

FLASH FORWARD TEN YEARS to an early-autumn evening when the leaves had just started turning color. I was paged by the hospital to the NICU. Rebecca, the nurse on call, met me at the sink and briefed me on two preemies who had been admitted a few days before. Fraternal twins, Jordan and Emily were born at just twenty-five weeks of age—more than three months before their official due date. I put on a clean gown, blue nitrile gloves, and a mask, as the last thing these babies needed was to be exposed to anything I may have inadver-

---

*We used to think that the mitigating factor explaining the difference in longevity between the sexes was behavioral in nature. More men, for example, have typically perished while serving as soldiers, and while employed in more dangerous occupations. We now know that genetic females' longevity advantage can be attributed to factors that are biological in nature.

tently tracked in from the hospital atrium, where I'd been sitting just a few minutes before my pager went off.

Rebecca had been on staff with the hospital for more than three decades, and despite the long hours and the very difficult work in the NICU, she looked much younger than her sixty-odd years. Rebecca had one of those voices and ways of being that left you feeling reassured, no matter how dire the situation. Most of the staff, including many of the physicians, often deferred to her when thinking about changing a medical care plan for the hospital's smallest patients. The senior nurse in the Level 4 NICU, Rebecca truly was a preemie whisperer. And what she said to me that night would change the course of not only my research but also my life.

Thankfully, most of us are not aware of the struggles that newborn premature babies must endure just to make it through the day. Tiny and frail, they have to fight to survive, alone in their diminutive translucent homes. Those incubators, crudely conceived artificial wombs, serve as their controlled environments until the babies are old and strong enough to not need them anymore.

A Level 4 NICU usually houses the youngest and sickest of the premature newborns. Many of the incubators used here have an air-filtration system that keeps the risk of infection lower by protecting the babies from the outside world. The incubators also produce the right amount of moisture in the air. When babies are born very early, their skin often hasn't finished forming yet and is unable to provide the proper barrier needed to avoid dehydration.

An immense amount of technology and human capital is invested in the precious few who occupy these Plexiglas enclosures. Nurses, doctors, and family members are caught up together in a constant struggle to help keep the babies alive, to encourage them to grow and thrive.

You never really get used to the sounds of the equipment in the NICU. The fans hum, the monitors buzz, and occasionally an alarm beeps loud enough to disorient even the most hardened of medical staff. It's no wonder research has shown that the light and sound spectacle of modern medicine can have a negative impact on the health of premature babies (it's something doctors are trying to fix these days).

My introduction to the NICU was hard and fast, first as a medical student and then as a physician. My time there oscillated between pure awe and utter terror, and I often experienced both emotions in quick succession—and sometimes simultaneously.

Mostly, though, there is a lot of waiting. With all the medical advancements we've made over the years, it's still time, more than anything, that these young bodies need. The babies are in the ultimate inverse of a race against time—their biology needs as much time as it can get to mature. They end up in the NICU for all sorts of reasons, of course, but in many cases they're there because a premature birth endangers the brain and lungs, which take longer to develop than the other organs do.

Often one of the biggest challenges for the youngest preemies, and one that determines their chances of survival, is the degree to which their lungs have developed. The lungs of premature babies have to acquire oxygen and release carbon dioxide at a rate that is compatible with life long before they are meant to. We're still not sure why some babies are born prematurely, but thankfully, over time, we've developed improved interventions to increase their odds of survival.

Regulating body temperature, as well as keeping in check the trillions of microbes always on the lookout for an easy target, can prove to be too much work for some preemies. It's a miracle that these babies, separated from the protective envelopment of the womb long before they're ready to face external challenges, can survive months before their due dates. But survive they do. All sorts of things can ultimately contribute to the death or life of a premature baby—from gestational age at birth to unanticipated bumps along the road. And surprisingly, one of the most important indicators of potential success in dealing with the rough-and-tumble adversities of life comes down to one simple thing, as I was just about to discover.

After I examined Jordan and Emily, Rebecca led me down a long hallway and into a quiet room where I could spend some time with their parents. Hospitals often lack the physical space for concerned family to congregate comfortably. We were lucky to have a room for talks like this.

I sat down with Sandra and Thomas to discuss our care plan for their twins, but before too long they were telling me about their journey to parenthood. After so many failed attempts, numerous rounds of hormone injections, and even in vitro fertilization, they had all but given up on having children of their own.

And then it finally happened. They were overjoyed at finding themselves pregnant, but tried not to get too excited at first. They knew from personal experience how quickly good news could turn to bad. But as the days and weeks went by, they gradually allowed themselves to believe that this pregnancy might actually lead to happiness. When the sonogram showed that Sandra and Thomas were pregnant with not only one child but two, their dream of having a family at last seemed to be coming true.

And just when they let themselves take a breath, trouble struck again. They went from imagining what it was going to be like to have their quiet Brooklyn apartment filled with the lively sounds of two young children to hoping and praying that their twins would survive.

Rebecca had me paged late one night because she wasn't happy with how Jordan was looking. Her years of experience had taught her this: her instincts were almost always right. Having taken care of the twins since their admission, I found myself looking forward to seeing them—they had been changing so quickly from the first day they were admitted. So this news from Rebecca was upsetting. It was true that after two weeks in the NICU, Emily and Jordan had been thankfully breathing well on their own, but I knew that they weren't out of danger yet.

On my way to Jordan's incubator, I tried not to get tangled in all the wires hooked up to the machines that were helping this child along. Rebecca, after having gone through the same routine that I did every time I went in, without fail—washing hands, gowning, donning gloves and mask—met me at his bedside. We both knew that things can be precarious for patients this young. Rebecca warned me then that I should prepare for the worst in Jordan's case. And she was right. Twelve hours later, Jordan passed away.

I ran into Rebecca a few years later, this time in the hospital cafeteria. I had moved to a different institution and had come back to give a lecture. After so many years of devoted service, Rebecca was getting ready for retirement at the end of the month and looking forward to spending more time with her own seven grandchildren and two great-grandchildren. I told her that my experience with her in the NICU that night was still very fresh in my mind.

"Yes, they never leave you," she said. "I still remember every one of their faces." She reached for her coffee to take a sip.

"There's something I've been meaning to ask you," I said to her. "That night in the NICU—how did you know about Jordan? What was it that made you think he wasn't likely to make it?"

"I'm not sure . . . but once you do this job for so long, you develop a feel for things. And so much of what we do is a judgment call. Sometimes it's even something that the lab results or testing don't always show you initially. Maybe it's just intuition. One thing's for sure, though: in the NICU, it's almost always so much harder for boys than for girls. And I guess it's not just in the NICU . . . I lost my husband twelve years ago now, and most of my girlfriends are widows too."

I was quiet while I reflected on what Rebecca had just shared with me. I couldn't help thinking about my grandmother and the dearth of men at the far end of the human aging trajectory. It was as if everything I had ever researched and experienced clinically was coming together at that moment, forming a crisp question out of the fog of years of study.

"Males, I was always taught, are the stronger sex. Yet that's the opposite of what I've seen so far, both clinically and in my genetics research. So why do males seem like the weaker sex in fact?" I asked.

"Maybe you're just not asking the right question," she said thoughtfully, stirring the remaining coffee in her cup. "Instead of thinking about male weakness, maybe the question you want to ask is, What makes females stronger?"

THE ANSWER TO REBECCA'S QUESTION came to me six years later: It was a beautiful summer's day—perfect for a drive down to the beach. The sun was finally out after a very long winter and a very wet spring. I promised my wife, Emma, some quiet time for just the two of us, and as I wasn't on call that day, I even turned off my phone. The last thing I remember was reaching over and holding her hand as we were driving westbound on a mostly empty street, singing along with the song that we heard the first time we ever danced together, Leonard Cohen's "Dance Me to the End of Love."

Witnesses later told us that we were hit dead-on broadside

by someone who ran a red light and barreled toward us at more than forty-five miles per hour. Our car rolled twice. The impact was severe, the roof of our car caved in, and none of the airbags deployed. Because of the extent of damage to our car, the first responders were preparing themselves for horrific traumatic injuries. We were lucky to be alive.

We both had some bruising and bleeding from the tempered glass that shattered and rained down on us when the car flipped. Given what we had just experienced, our injuries turned out to be pretty minor and pretty similar—but Emma's injuries were a bit more serious. So, you know what I was thinking while strapped to a spine board in the back of an ambulance hurtling toward the hospital? I was thinking about how grateful I was that Emma was a genetic female with two X chromosomes.

I thought back to when Rebecca had suggested that I ask myself why women are stronger at both the beginning and the end of life. I knew from my clinical work and research that even if my wife's injuries were the same as mine, given the odds, she was likely to make a better and faster recovery than I was. Her wounds would heal faster, and she would have less of a chance of subsequent infections because of her superior immune system. All in all, her prognosis was almost assured to be better than mine.

This was because her body had the use of two X chromosomes, while mine had the use of only one. To review the basic chromosomal differences between the sexes, the cells of all genetic females have two X chromosomes, while those of genetic males have one X chromosome and one Y chromosome.* When it comes to dealing with the trauma of life, genetic females have options. And genetic males don't.

Our two sex chromosomes are given to us by our biological parents before we are born. My wife's genetic superiority began long before we ever met. When she was only twenty weeks old within her mother's womb, she already had a survival advantage over me—one that will continue at every data point throughout our life cycles. That's even if we adjust for other lifestyle and behavioral risk factors such as occupational hazards and suicide. From the beginning, she is likely to live longer than me no matter what life throws our way.

And my wife doesn't win only when it comes to overall longevity. Her risk for developing cancers in organs we both have, for example, is lower than mine. And if she does develop cancer, she has better odds of surviving, as research shows that women respond better than men to treatments. More women of course do develop breast cancer, but overall men still die of cancer in higher numbers per year than women.

The cost women seem to pay for having a more aggressive immune system, one that's better at battling both invading microbes and malignant cells, is being self-critical—immunologically speaking. The immune systems of genetic females are much more likely to attack themselves, which is what occurs in diseases like lupus and multiple sclerosis. So the only thing that I

---

*Most humans inherit two sex chromosomes, written formally as 46,XX and 46,XY. It's possible to inherit many other variations of the sex chromosomes, including 45,XO, 47,XXX, 47,XXY, and 47,XYY.

have going for me is a lower chance of developing an autoimmune condition.

What I knew that night as we rushed to the hospital was that
my wife's cells were already starting to divide, undergoing a process of cellular selection to deal with the microbes that likely entered her body on impact. They were already starting to draw on
their collective genetic wisdom to undertake repair work on her
tissues. And in each area of her body, be it her leukocytes, which
are part of her immune system, or the epithelial cells that make
up her skin, her cells would be going through an empowered and
flexible genetic process of selection. My body, being composed of
cells that are genetically identical, didn't have this option.

Although every genetic female has two X chromosomes in
each of her cells, every cell has the use of only one. My wife's
cells use either the X chromosome she inherited from her
father or the one from her mother. My cells don't have that
luxury. Every one of my cells has to use the same exact X chromosome, the one I inherited from my mother, while my Y
chromosome couldn't do that much after the accident we
experienced but sit there watching helplessly.

The ability to use different X chromosomes is one of the
main reasons for my wife's genetic superiority. As our room
filled with GET WELL SOON balloons, cells in her body were using different Xs and continuing to rapidly divide. What began
as a fifty-fifty split between cells that were using the X from
her mother and those that were using the X from her father was
now rapidly skewing toward using one particular X, whichever
one happened to be more effective at doing the job required.

Even before the first emergency responders arrived, more of those white blood cells were dividing, using one X over the other. And to deal with the challenge of healing, the same cellular competition of using only the best X for the job was likely happening everywhere else in her body as well. If you'd looked inside *my* blood expecting to find the same thing, you'd have been disappointed.

Having the use of two X chromosomes makes females more genetically diverse. And the ability to rely on that diverse genetic knowledge is why females always come out on top. Whether it's an infant girl's survival in the NICU, a woman's ability to fight infections, or a genetic female's decreased risk of having an X-linked intellectual disability, it all boils down to the simple fact that females have a degree of genetic flexibility that males lack.

Although we belong to the same species and are more similar than we are different, there's an important reason that females are more genetically endowed. Our very existence has depended on it for millions of years. Being the stronger sex, genetically speaking, is what allowed females to survive long enough to ensure the survival of our offspring—which in turn means the survival of us all.

My original genetics research and clinical discoveries, my life experiences, the groundbreaking work of my colleagues, and the findings of pioneering scientists challenging the doctrines of their time have culminated in this undeniable understanding: women are the stronger sex.

In *The Better Half*, I will explore the key challenges that occur throughout life and show how genetic females conquer them, leaving males behind when it comes to longevity, resilience, intellect, and stamina. I will tackle how medicine and pretty much everything else have all dismissed this fact.

When I was in medical school, I learned to expect that more of my female patients would report myriad side effects from the medications I would prescribe. I was also taught that the likely reason for this was behavioral—that women were just more vocal about any problems and generally saw their physicians more often than men did.

But if it's just a reporting bias, then why are so many women experiencing serious side effects that require significant medical intervention? A U.S. General Accounting Office review of ten drugs taken off the market revealed that eight of them were withdrawn because they were found to be dangerous to women. In addition, it's women who are more frequently overdosed unintentionally by the doctors aiming to treat them.

Although we've known for years from a medical perspective that women are more sensitive to chemical compounds such as alcohol, we still, for the most part, prescribe drugs to both genetic sexes as if they're exactly the same. This needs to change. Almost twenty years ago the Institute of Medicine of the National Academy of Sciences published a report that claimed the following: "Being male or female is an important fundamental variable that should be considered." So let's consider it.

Outside obstetrics and gynecology, the incredible advances in modern medicine that we're all benefiting from have been almost entirely produced from research that exclusively used male participants, male research animals, and male tissue and cells. The chasms created by the lack of female research animals and female tissue and cells in preclinical drug trials leave most physicians having to estimate or at worst outright guess what might be the best dose or treatment for their genetic female patients.

When I was designing studies to test the microbe-killing power of an antibiotic I discovered almost twenty years ago, I remember how naive I was regarding the inclusion of women in basic and clinical research. To further test the effects of one of the drugs I discovered, I contacted a company that specialized in running experiments independently, so that I could corroborate or refute my findings. While designing the studies for the company to perform on my behalf, I assumed that it would be using an equal number of male and female mice.

I was wrong. It used only male mice. As I came to learn, that company wasn't alone in this. Everyone else was doing exactly the same thing. When I inquired as to why, I was told at the time that it was easier (and cheaper) to use males. Interestingly, as I was to discover, female mice can have much stronger immune systems, which could complicate the results of an experiment that's trying to cure infections equally in both sexes.

Indeed, we've misunderstood the physical abilities of women and discounted their genetic strength for far too long. In *The Better Half*, I will outline how our perceptions, health care, and research culture need to change. The future of medicine and our survival as a species depend on it.

1

THE FACTS OF LIFE

THIS IS A BOOK ABOUT CHOICES. Not the choices we consciously make every day, but the biological ones that happen every single second in every single genetic female. This phenomenon starts the instant a mother's egg accepts a father's sperm and the process of fertilization begins.

Here's some basic biology that will be required as I develop my argument: Each human cell has a total of forty-six chromosomes. Two of those are the sex chromosomes; and if you've inherited an XX pair, then you're a genetic female, and if you've inherited an XY pair, then you're a genetic male.*

Like an instructional set of encyclopedias, our twenty-three pairs of chromosomes have within them genes that provide the genetic information that makes our lives possible. It's thought that we have a total of about twenty thousand genes that are spread across our twenty-three pairs of chromosomes. While some of our chromosomes contain more genes than others, every one of our chromosomes is as significant as the other.

For the most part, each of the twenty-three chromosome pairs shares versions of the same genes, except if you're a genetic male and have inherited an X and a Y. The X chromosome contains almost one thousand genes, but the Y chromosome has only about seventy, most of which are involved in making sperm.† For many years it was thought that one of the genes on the Y chromosome was also responsible for the excess hair found on the ears of males as they get older, the medical term for this being "auricular hypertrichosis." More recent studies have suggested that even this unglamorous ability can no longer be attributed solely to the Y chromosome.

Even without understanding all the scientific processes

---

*As I mentioned in the introduction, many variations of sex chromosomes can be inherited. These include such rare variations as 45,XO, which is called Turner syndrome; 47,XXX, which is called triple X syndrome; 47,XXY, which is called Klinefelter syndrome; 47,XYY, which is called Jacob's syndrome; 48,XXXX, which is called tetrasomy X; and 49,XXXXX, which is called pentasomy X.

†Recent research is uncovering new health implications for many of the genes found on the Y chromosome. Unfortunately for genetic males, most of the news has not been positive. Some of the genetic information on the Y chromosome has been implicated in everything from increased inflammation, to a suppression of the protective adaptive response of the immune system, and even to an increased risk of coronary artery disease.

that occur during conception, we've reached a point in our evolution as a species when sexual intercourse is no longer required to make a baby. We are well on our way to mastering the arts of manipulative conception. Once the stuff of science fiction, assisted reproductive technologies—which can fertilize human eggs outside the body under extremely sterile laboratory conditions—are now commonplace. But we are also capable of so much more than that. We can create a child using the genetic and cellular material from three different parents and even edit our own DNA.

Nevertheless, the so-called "natural" process is anything but simple. Around five hundred million sperm start the journey to the egg, swimming up a mother's reproductive tract at a remarkable speed. They move through her cervix and uterus and eventually into one of her two fallopian tubes. There they meet an egg. And whether the single sperm that successfully bores its way through the outer layers of that egg is carrying an X or a Y chromosome is what sparks a genetic course that will determine your biological fate. Everything from your lifetime risk of developing cancer or a neurological condition, like Alzheimer's disease, to your ability to fight off viral infections is decided in that very moment, depending on which set of sex chromosomes you inherited: XX (female) or XY (male).

Your biological sex isn't always the same as your gender. Gender is dependent not on your sex chromosomes, but rather on your sense of maleness, femaleness, or anything in-between or beyond. Gender is an individual's self-concept and self-identification, as well as the role an individual may assume in society. Gender is often assigned to children at birth based on sex chromosomes and external genital anatomy. This can even happen long before the time of birth, after, say, a sonogram or the chromosomal testing of a fetus from a blood test on the mother.

Individuals may fluidly accept or change their gender, which may not align with the one assumed at any point throughout the life course. Yet when it comes to sex chromosomes and their immense effects on our lives, there's no choice. An individual doesn't choose to inherit a Y chromosome or two X chromosomes or any combination of both.

When it comes to sex differentiation in humans, genetic variations can occur within genes that alter the course of physical development. The *SRY* gene that is found on only the Y chromosome is a crucial player in sexual differentiation, as it triggers the process of creating testes out of the bipotential gonads in the fetus, which then secrete testosterone. This cascade of cellular development triggered by the *SRY* gene also results in the development of the male external genitalia. But if the cells of a person with an X and a Y chromosome cannot respond to testosterone, then in these rare cases they will develop externally as females but internally with testes, and no uterus, fallopian tubes, or cervix. This is exactly what happens in the case of complete androgen insensitivity syndrome (CAIS), a genetic condition that results from a mutation that's inherited in the androgen receptor, or *AR* gene. Most of these XY individuals do not know that they have CAIS until puberty begins and they don't start menstruating.

Rarely, a baby born with two X chromosomes can develop physically along a genetic male developmental pathway. This can happen when a small part of the Y chromosome that houses the SRY gene is inherited along with two X chromosomes. Although uncommon, it's possible for a child to develop externally and internally as a male without the SRY gene or any part of the Y chromosome at all. I was involved in the discovery of an exceptionally rare alternative pathway of sex development in a boy named Ethan, who was born a biological male, with two X chromosomes and no SRY gene or any other genetic factors that cause sex reversal—something that was not thought to be genetically possible. What we discovered was that Ethan had a duplication of the gene SOX3 that in his case turned a genetic XX female into a physical male. The SOX3 gene is thought to be the genetic predecessor of SRY, and they both play a crucial role in sex differentiation.

Human sex development is complex. Even today, geneticists and developmental biologists are still trying to untangle the seemingly endless pathways of sex differentiation. We do know that chromosomal sex and the differences based on it are biological. And here's why: human female eggs contain only one X chromosome, and therefore it is the male sperm that will determine the biological sex of the child. If a sperm is carrying a Y chromosome, more often than not, a genetic boy will eventually develop. And all the cells within that boy will have to use the same identical X chromosome—the one he inherited from his mother. On the other hand, if that sperm is carrying an X chromosome, then the fertilized egg will develop along a preprogrammed genetic pathway into a female.

For most of human history we simply didn't know how the sex of a child was determined—or, at least, we didn't have the tools to prove, with scientific certainty, how sex differentiated. There were many theories, and respected members of many cultures relied on signs from the gods or elaborate lunar calendars. Some people in India continue to trust in ancient Ayurvedic remedies in an attempt to ensure the birth of a boy. I've even been told by some religious women that they have been encouraged to focus on images of saints while having sex in order to maximize the likelihood of conceiving a saintly son.

Historically, the importance of having a male child (especially in a patriarchal society where the inheritance of position and property was transferred solely through a male heir) drove people to try almost anything to sway the outcome in favor of XY. More than two thousand years ago, Aristotle turned his attention to this problem, likely at the behest of some older male patrons who wanted to ensure that they would get a male heir. He was already fascinated with the embryological origins of animals and had become an avid collector and dissector of any embryos he came across. Especially plentiful, given their relative size and easy access, were the fertilized eggs of the domesticated common fowl—a.k.a. the chicken.

Aristotle documented his findings within the pages of *The Generation of Animals*, published in the middle of the fourth century B.C.E. In it, Aristotle accurately describes by today's scientific standards some of the variations at the start of life. He correctly theorized that some animals (like the chickens he was dissecting) are born from eggs, while mammals with placentas enter this world through a live birth, while still others like sharks can have eggs that actually hatch inside them. Aristotle is thought to be the first person to figure out the purpose of the placenta and umbilical cord.

But when it comes to how male and female development diverges, Aristotle's theories haven't all aged well. He posits that it's the amount of *heat* provided by the male partner during intercourse that later determines the sex of the child. A specific amount of heat was thought to be a necessary energetic substance for all babies to develop. The more heat that was provided to the embryo by the father, the more likely the embryo would later develop into a boy. Not enough heat and your child would turn into a girl. Females were, after all, thought by the ruling men of the time to be half-baked males. The greater the heat of the fires stoked by passion, the more likely a woman was to give birth to a boy.

What to do, then, when there just wasn't enough passion in the moment, or the man was perhaps too old to get really excited but still wanted to produce a male heir? Aristotle's solution was simple: the couple should try to conceive during the warmer months of the year, with summer, of course, being ideal. Before this is dismissed as pure quackery, Aristotle was actually on to something when he thought that "heat" played a role in determining the sex of a child. Just not in humans.

When it comes to some vertebrates, such as alligators, turtles, and some lizards, the incubation temperature of their fertilized eggs can influence the sex of their babies. Higher temperatures can favor male development in crocodiles as well as in the living fossil tuatara, a reptile endemic to New Zealand. But for many other species of vertebrates, such as the European pond turtle and the spur-thighed tortoise, higher temperatures during incubation actually leads to females.

The idea of "baking" a male persisted for a long time and was even adopted by the early Christian Church. It might be difficult to believe, but there are still people who think that exposing a woman to a lot of heat—not only during conception but also throughout the pregnancy—increases the odds of having a boy.

I first heard about the beliefs surrounding heat during pregnancy from a pregnant patient named Anna. With three girls already at home, and her partner the only son in his family, Anna was holding out hope that this fourth child would be a boy.

When I met her, she told me that she wasn't enjoying the pregnancy very much. Anna was under an incredible amount of pressure. Her mother-in-law was a big believer that more heat would produce a male, and she procured for her daughter-in-law an Ayurvedic medicine that was meant to raise her internal body temperature.

Unfortunately, pregnancy and many of these herbal medicines just don't mix, even if the tinctures and teas are naturally derived. Anna indeed delivered a boy a few months later. He had multiple congenital anomalies, which very likely were caused by the teratogenic effects of the elixir she was drinking.

More than a thousand years after Aristotle, as the science of medicine (led almost entirely by men) advanced its understanding of numerous important phenomena—from the English physician William Harvey's description of the circulation of blood in the seventeenth century to Edward Jenner's early use of a vaccine against smallpox in the eighteenth century and even the Nobel Prize winner Wilhelm Conrad Röntgen's discovery and use of X-ray imaging in the late nineteenth century—there was still no scientific consensus about how the sexes were determined. Indeed, most of the genetic history of both men and women has been written and rewritten by men, and this has, in my view, negatively impacted how we have come to treat both sexes from a medical perspective.

After nothing but narrow views on the origins of and differences between men and women, our understanding of the chromosomal underpinnings of sex finally began to take shape in the early twentieth century, and it happened as a result of discoveries made by pioneering women scientists. One of them was Nettie Stevens.

As Stevens was studying the chromosomes of the mealworm, she discovered what had eluded others for so long. Her mealworms revealed that both the female and the male of the species had twenty chromosomes (as you'll recall, humans, on the other hand, have forty-six chromosomes in total). But in males, one of the twenty chromosomes was a lot smaller than the others. What Stevens had discovered was the Y chromosome.

In her landmark paper in 1905, Stevens postulated and then described chromosomal sex determination. Her work spelled out for the first time that females have an XX sex chromosome complement while males have an XY. She figured that it was this difference that split the sexes down their unique paths of development.

During college I never learned about Stevens. I was told that our understanding of sex chromosomes was due to someone else: Edmund Beecher Wilson, a contemporary of Stevens's and a senior geneticist who was touted as the originator of the concept of the sex chromosomal determination system. What my textbook neglected to mention was that Beecher had access to Stevens's research results before they were published. In addition, his paper (now with similar results to Stevens's) was rushed through publication in the August 1905 *Journal of Experimental Zoology*, a journal where Beecher just so happened to serve on the editorial board.

Another female scientist who doesn't always get her due is the English geneticist Mary F. Lyon. Her research was critical and deserves attention during our brief primer. Lyon rocked the genetics world when she published a paper in *Nature* in 1961. In just a single page she changed our thinking and understanding of genetics forever, and the implications of her hypotheses and findings are still being researched and studied today. Through her work studying coat color in mice,

Lyon provided the basis for our understanding of how men and women differ genetically. She described "X chromosome inactivation," meaning that one of the two X chromosomes in female cells "randomly" inactivates and is silenced during the earliest period of development—before a mother even realizes that she's pregnant.

What's amazing is that it's been more than fifty years since Lyon's visionary paper, and the truth is that we still don't really understand all the steps involved in X chromosome inactivation or silencing. How does a cell choose between two X chromosomes at the beginning of life? Is it a competition? How is X inactivation suppressed in genetic men who are XY?

One of the challenges is that this enigmatic process is hidden from view. We think it happens at around the point in development when there are only twenty cells, after the egg first embeds in the uterus. One way to solve this enigma would be to work with human embryos in vivo, but that comes with its own ethical challenges.

At this very early stage of pregnancy, the group of cells that will eventually form a baby already has a chromosomal sex, either XX or XY. Yet it's only within each of the XX female cells that a process of X inactivation begins to occur. And the XX female cells do all of their X inactivating in the uterus, far from prying scientific eyes. Which is why, when it comes to X inactivation in human cells, there's still a lot we don't know.

What we do know is that human cells use an RNA gene

called *X inactive specific transcript*, or *XIST*, which is found on the X chromosome and produces a scaffolding that covers the soon-to-be-silenced X chromosome from top to bottom. During this early phase of development, both X chromosomes are not silenced, and they both express *XIST*, but only one of them will eventually be subdued and silenced. Since male cells do not normally have more than one X chromosome, the process of X inactivation doesn't have to occur within them.

So which of the two female X chromosomes becomes silenced? For the most part, the superior one of the two outwits *XIST* and stays active. I have had female patients, for example, who had an X chromosome that was damaged or abnormal, and within their cells this damaged X chromosome was always the one that was preferentially silenced and turned off. The *XIST* scaffolding works by squeezing and eventually condensing down and silencing the X chromosome into a structure called a Barr body.* Every female cell ends up having one active X chromosome and one silenced X chromosome in the form of a Barr body.

As in a good mixed martial arts (MMA) fight, only one X is left standing in each cell. If each of the X chromosomes is equal in this X inactivation match-up, then the silencing is thought to be random—like the result of a coin toss. This

---

*Up until recently we knew of only two small regions on the tips of the X chromosome, called the pseudoautosomal regions 1 and 2, that are still active on the silenced X or Barr bodies in females. These genetic regions are very small—containing only about thirty genes, or just a few pages of genetic material—when compared with the voluminous rest of the X chromosome.

process ends with the silenced X chromosome or Barr body becoming inaccessible to the female cell. Or so we thought.

For most of the fifty years since Lyon's paper about X inactivation, we assumed that the genetic machinery of a woman's cell was not able to "open," or access, the Barr body (remember, this is the silenced X). It turns out that Lyon wasn't 100 percent right: the silenced X isn't *completely* turned off. Rather, women have the use of two X chromosomes in every one of their trillions of cells—the silenced X is still helping the cell survive. About a quarter of the genes on the "silenced" X chromosome are in fact still active and accessible to female cells. We call this phenomenon "escape from X inactivation."

As I'll show in subsequent chapters, having another X chromosome provides extra genetic horsepower to each cell, which is an advantage that females have over males. The fact of the matter is this: If you are a woman and have inherited two X chromosomes in each of your cells, like the three and a half billion other genetic females on this planet, your cells have options. And when the going gets tough in life, those options help you survive.

Like each volume in the genomic encyclopedia set that I mentioned earlier, every chromosome provides genetic instructions from which we draw on every day of our lives. Need some more pancreatic lipase to help you break down the fat in that pistachio gelato you just ate? No problem. The cells in your pancreas will use the instructions from the *PNLIP* gene found on chromosome 10 to make some more. What about the lactose in that gelato? No problem again. Cells that line your gut can rely on instructions from the *LCT* gene that's found within chromosome 2 to make all the lactase (the enzyme that breaks down lactose, the sugar in milk) you need to keep from feeling bloated.

So why is the X chromosome in particular so important in the game of life? Well, without it, human life isn't possible. No one has ever been born without at least one X chromosome. Besides making life possible, it also provides us with a foundation from which we build and maintain a brain and from which we create our immune system. It's a rich genetic volume of instructions that orchestrates the development and many crucial functions of the human body.

HUMANS ARE NOT the only creatures on earth that use their chromosomes for sex determination. I first started working with honey bees over twenty years ago, and my research interests were initially sparked by a very simple question: What happens to a honey bee when it gets sick?

Honey bees have to collect pollen and nectar from numerous flower sources, often far away from their hives. And along this journey, they are exposed to all types of microbes.

Unlike vertebrate animals such as humans, honey bees don't make antibody proteins to fight off an infection when they've been invaded by a microbe. Instead these insects have become quite adept at chemical warfare. Like a personal pharmacy on demand, honey bees are able to custom-make their own signature antibiotics to treat themselves when they have a microbial infection. (Some of these antibiotics, like apidaecin, can even sneak their way into the honey we consume.) The goal of my research with bees was to discover whether we could use the antibiotics that honey bees make to treat humans who have infections.

As a geneticist, I was fascinated by honey bee reproduction and genetics. Unlike many other animals, such as birds that use something akin to the XX and XY system, honey bees have a unique way of determining sex. I was reminded of this when I opened a hive one day and noticed the eggs the queen bee was actively laying in this hive—and she was a prodigious egg layer indeed. Queen bees lay eggs at a rate of about 1,500 per day.

Unlike Aristotle's patrons who would do anything to have a say in the sex determination of their offspring, queen bees mastered the art of sex selection millions of years ago. The queen bee herself can make the royal decision of whether to lay an egg that will turn into a female worker bee or a male drone bee.

Here's how it works: The queen lays an egg that has sixteen chromosomes, and if she does nothing more, it will develop into a male drone bee.* But if a queen wants to make

*In a healthy hive, in the middle of a productive summer, drones can make up about 1 to 15 percent of the fifty thousand to seventy-five thousand bees in the hive's overall population.

a female worker bee, she adds a little dollop of sperm, which has been stored in her body, onto the egg. This sperm mixes with the egg and fertilizes it. The sperm that fertilized the egg adds another sixteen chromosomes for a total of thirty-two. That's how many chromosomes it takes to make a female worker honey bee. While human females have an extra copy of an X chromosome, female honey bees have even more genetic options at their disposal. Every one of those sixteen extra chromosomes allows female honey bees to have more genetic choices than their male counterparts do.

Imagine that for a moment. Unlike human females, who have only one extra X chromosome compared with males, their honey bee compatriots have an entire extra set. Given all the duties entrusted to a female worker bee, it's no wonder she has so much extra genetic material. For one, to ensure that the hive stays as germ free as possible, female honey bees spend an enormous amount of time and energy maintaining it. They also serve as guards, putting their lives at risk protecting the hive's entrance if it's threatened by predators.

Female honey bees are also entrusted to find all the nutritional sources the hive needs to survive. Then there's the astonishing conversion of nectar into honey, which requires days of intensive effort. The first step to making honey is to add enzymes to digest the nectar. To aid the process, the female worker bees' wings must buzz at a stroke rate of 11,400 times per minute. Their distinctive buzz is required to help dehydrate the liquid nectar, eventually turning it into honey.

With all our scientific advances to date, humans have not yet found a way to successfully replicate this process.

A female honey bee can eventually progress from cleaning duty, to guard duty, to leaving the hive in search of pollen and nectar. It takes about two million visits to flowers, requiring an overall flying distance of fifty-five thousand miles, to make a single pound of honey. Not to mention that in the collection process, while they are avoiding predators, female honey bees also manage to pollinate 80 percent of the fruits, vegetables, and seed crops in the United States alone. If that isn't enough, they also communicate to their fellow flying female workers through an elaborate dance, which lets them know where to find a good food source. Female honey bees have also been discovered to be the advanced mathematicians of the insect world. Australian and French researchers taught female bees how to do arithmetic operations such as addition and subtraction. This ability was thought to be out of reach for any insect, as it requires the capacity to perform complex cognitive processes. But not for a female honey bee.

So what's left over for a male drone bee to do? The answer is simple . . . nothing.

Drones don't maintain the hive, they can't produce food for themselves, and they're kept alive and clean by the female workers. They can't even help defend the hive; instead of the stinger that a female has, a male bee has a phallic structure to use for the only thing they're good for: sex.

The sperm that makes females comes from a mix of male

drone bees from another hive that have had sex, usually in midflight, with a queen bee. A queen's virginal midair sex flight happens only once in her life, during which she mates with as many as fifty males. She stores the sperm in a specialized organ called a spermatheca. Queen bees have been known to keep sperm alive within them for a few years, using it only when they want to make females.

It's no wonder that most of the male bees in a hive are kicked out just before winter. The female workers don't want to have to care for them over the harsh months. Most of these males don't last long outside the hive, eventually succumbing to starvation, exposure, or predation.

It's easy to see why genetic options are essential to a female bee's industrious and complex life. Female honey bees are the undisputed champions of their sexed world.

Back in the human world, the extra X chromosome in females gives them the advantage of genetic diversity, which helps them more effectively deal with the challenges of life. Women are the ultimate biological problem-solvers—they have more solutions in their genetic toolkit. If each X chromosome has about one thousand genes, that means women can rely on cells that use a different copy of each one of those genes.

Usually, it's not an exact backup but an entirely different version of each of those genes that's found on the X chromosome. Think of it this way: If you need any old screwdriver, sure, go ask a man to get it out of his genetic toolkit. But if you simultaneously need two specific screwdrivers—a

Phillips-head and a Robinson—you'd better ask a woman, because she'll have both.

—

DESPITE THE GENETIC SUPERIORITY OF FEMALES, fewer girls are born every year than boys. At first blush, it may not seem like a striking difference, but it's a phenomenon that's worthy of our attention. There are 105 boys born for every 100 girls in the United States, and the statistic is pretty much the same around the world. You might think this proves that men are the stronger sex, but it's because women are so much more difficult to build that there are fewer of them.

As Mary Lyon discovered, the cells that eventually give rise to a female embryo all have to go through a multifaceted initial-developmental process of partially shutting down and safely stowing one of their X chromosomes. As far as geneticists know, this may be one of the most sophisticated tasks that occur during development. And it's what allows female cells to choose between two chromosomes.

So, like diamonds that require an immense amount of pressure and energy to form one hundred miles under the surface of the earth,* women are just harder to make. (For that very reason, and just like diamonds, women are also harder to break, which I'll talk about in my chapters on resilience and stamina.)

---

*Diamonds that are much rarer and blue in color are thought to form even deeper, some four hundred miles under the surface of the earth.

What happens if X chromosome "silencing" doesn't work out as planned? What we know from research on other mammals is that if an X chromosome is not properly silenced and turned into a Barr body in all the initial cells of a genetic female, a pregnancy is unfortunately lost. There has never been a human born with two fully active X chromosomes in all their cells. And if both Xs are accidentally silenced, the pregnancy won't continue either. This is why more female embryos are lost during the earliest stages of pregnancy, often before a woman even knows she's pregnant.

The cells that make up a male embryo are much simpler. None of these cells has to be silenced or has to inactivate an X chromosome to turn it into a Barr body. That's why males are easier to make. They have only one X, after all.

When it comes to female genetic superiority, you may think that the story ends here. But this is actually where our story begins. Not only do women have more genetic options to choose from within each cell, but they also have the capacity to cooperate and share that diverse genetic knowledge *between* cells. This form of cellular cooperation is happening simultaneously between and across trillions of female cells as they work together, pooling their collective genetic wisdom to tackle obstacles.

That superior cellular cooperation creates the fertile ground for a unique resilience that only women possess.

2

RESILIENCE: WHY WOMEN ARE HARDER TO BREAK

D R. BARRY J. MARSHALL was getting desperate. A few years after he and the pathologist Dr. Robin Warren developed the theory that it was a microbe giving their patients peptic ulcers* and even stomach cancer, there was only one problem: they didn't have many supporters. The entrenched medical establishment during the early 1980s thought that it knew

---

*Today, peptic ulcer disease (PUD) is used to describe a sore or break in the mucous membrane lining of the gut. Although collectively the sores are referred to as PUD, when a sore is found in the stomach, it's called a gastric ulcer; when it's in the upper portion of the small intestine, it's called a duodenal ulcer.

better. Who did these unknown upstarts from the backwaters of Western Australia, with no substantial research or publication background, think they were?

Long before Drs. Warren and Marshall started to work together in 1981, it was assumed by medical experts that gastritis and peptic ulcers were the result of too much stress and an improper diet that included spicy foods. And for the most part, no one questioned this well-entrenched dogma. The common treatment at the time was the use of histamine H_2-receptor blockers, a family of medications that reduced the amount of acid produced in the stomach. This was a rational choice, given the idea that ulcers were caused by behaviors that led to an overproduction of acid. Far from challenging the "too much acid" dogma, surgeons embraced this reasoning and started specializing in removing parts of their patients' stomachs and upper intestines as an easy way to make their life more manageable. For some reason, most of the patients affected were men.

Except every time that Warren found himself looking down his pathology microscope at biopsy samples from patients with ulcers, he was seeing something that went against everything that he was taught. "I preferred to believe my eyes, not the medical textbooks or the medical fraternity," wrote Warren in the book *Helicobacter Pioneers*. His initial work indicated that a screw-shaped microbe called *Helicobacter pylori* was the actual cause of the ulcers and stomach cancers that he was seeing in his almost exclusively male patients. The orthodox medical teaching at the time was that the stomach was simply too acidic for any bacteria to survive and grow, so the likelihood that anyone in medicine would take the doctors' findings seriously was infinitesimally small. According to Warren, "In fact, there was only one doctor who did believe in what I was doing . . . my wife Win, who was a psychiatrist and who encouraged me."

Warren and Marshall knew that they weren't imagining things every time they looked into their microscopes. They were sure that this microbe was living comfortably within the acidic environment—actually thriving, in fact. They reasoned that the bacteria could be causing inflammation in the lining of the stomach, which would eventually cause it to erode.

Ulcers, then, had nothing to do with stress or diet. What we needed to do was treat the pathogen. Kill the microbe, cure the disease.

As Marshall said in an interview in 2010, "I had this discovery that could undermine a $3 billion industry." He was of course referring to the pharmaceutical industry. And that was true: people were making vast sums of money off the idea that ulcers were caused by smoking, drinking, feeling the stresses of life, and eating spicy food. It shouldn't be too difficult to imagine why there would be resistance from people who were developing and selling entirely new classes of pharmaceutical drugs that lowered the acid in the stomach, making the pain from an ulcer bearable. But these drugs weren't really curing patients; they were just easing symptoms.

So, what's a young physician to do when he's growing tired of being ignored or dismissed? What can you do if both the medical orthodoxy and Big Pharma are trying to sweep your ideas under the rug? Dr. Marshall drank a foamy brown broth full of the microbe *H. pylori* extracted from one of his ill patients and hoped that it would make him sick.

And it did. At first, Dr. Marshall experienced some minor stomach discomfort, but by day five he was vomiting, and by day ten his stomach was inflamed, completely colonized by *H. pylori*. Inflammation and gastritis took hold of his body from the *H. pylori* infection, and he was well on his way to developing a full-blown ulcer. His wife, Adrienne, stepped in and convinced him that it was time to cure himself with antibiotics.

The antibiotics eradicated the *H. pylori* in Dr. Marshall's stomach, and he made a full recovery. But despite the additional experimental evidence supporting his theory, many clinicians at the time remained unconvinced. It took almost another ten years of scientific advocacy before clinicians began to take the theory seriously. Marshall and Warren not only went on to convince the world of the veracity of their theory but also were awarded the Nobel Prize in Physiology or Medicine in 2005 for their groundbreaking discovery.

Lucky for Dr. Marshall, he didn't try to infect his wife, Adrienne, with *H. pylori*. Millions of people around the world might still be suffering needlessly today if she had volunteered for this experiment, and he wouldn't have a Nobel Prize to celebrate. If she—or any other genetic female, for that matter—had drunk the broth instead, his experiment could have failed.

We've known for a very long while that ulcers were up to four times more common in men, though we didn't know why. But now, there is no doubt that it's because males don't have the same ability to appropriately fight off microbes, such as viruses and bacteria, that females do. Males are not only incapable of mounting the same vigorous immunological response against microbes as genetic females, but they are also more likely to get gastritis, peptic ulcers, and even gastric cancer as a result of them.

Recent research indicates that the differential responses men and women have to being infected with *H. pylori* may be mediated through hormones such as estrogens. In fact, giving male mice a type of estrogen called estradiol reduced the severity of gastric lesions caused by *H. pylori*. In humans, treating cell lines of human gastric adenocarcinoma (otherwise known as stomach cancer) with estradiol, for example, seems to inhibit their growth. So it may not be *H. pylori* alone that's the sole reason more men get cancer from being infected. Genetic females are just more resilient than men in dealing with the stress of the infection.

Differences in the level of the sex hormones such as estrogens and testosterone in our bodies are dictated by the chromosomes we inherit. The gonads that produce the sex hormones, such as the testes and ovaries, were formed depending on which sex chromosomes you have. If you've inherited a Y chromosome and your gonads are testes, then you'll have more testosterone than estrogens in your body. Without the Y, there will be more estrogens within your blood. Some female resilience is the result of sex hormones while other female resilience has to do with female chromosomal diversity, cooperation, and of course subsequent superiority as a result of having more than one X chromosome to choose from.

When it comes to resilience—overcoming life's trials and tribulations—genetic options help women confront one of the biggest challenges of all: not to be consumed by pathogens like the one Barry Marshall and Robin Warren discovered causes peptic ulcers.

The trillions of microbes in our environment are always on the lookout for an easy target, which is why every higher-order organism—be it an oak tree, a French bulldog, or a human being—has some type of immunological defense system. Skin and digestive antimicrobial enzymes provide important barriers that reduce the chance of entry or colonization by pathogens. But what happens when a physical barrier isn't enough?

Cue the immune system. It evolved to deal not only with pathogenic and cancerous rogue cells but also with parasites such as intestinal worms. The immune system isn't a distinct organ, like a heart or brain. And that's a good thing, because it needs to be active both spatially and temporally—everywhere, at any time, all the time.

When it comes to the overall differences between men and women and their ability to fight a multitude of microbial infections, the clinical outcomes are striking. Whether it is bacteria such as *Staphylococcus aureus*; or *Treponema pallidum*, which causes syphilis; or *Vibrio vulnificus*, which causes vibriosis, women are consistently better at fighting these infectious microbes.

Without a strong immune system, you might find yourself with an *H. pylori* infection you can't fight off, or something far deadlier. And it's not just bacteria that females are better at fighting off. It's viruses as well.

THE RAIN WAS DEAFENING and I could see the water levels starting to rise outside my window at the Tarn Nam Jai Orphanage. Eventually the entire street was flooded, and the children were cut off from the rest of the city.

There's a reason that Bangkok used to be called the Venice of the East. Long before all the canals were paved over, they were used to effectively transport people, animals, and goods. But on days like this one, in the middle of the rainy season of 1997, it was as though the past became present, and when the city began to flood, its side streets were submerged once more.

As the water levels continued to rise, I didn't have much time to reflect on Bangkok's past glory. There were a dozen children to tend to, some of whom were HIV positive. When your immune system has been decimated by a virus whose evolutionary strategy evolved to do just that, you need all the medical help you can get.

The problem with flooding is often not just the water but also what it carries with it. I noticed a nervous rat running in circles on a small wooden plank floating along the street. It was a sign that the sewers were mixing with what was quickly becoming a river in front of the orphanage. For the six children at Tarn Nam Jai who were HIV positive, exposure to a greater microbial load than normal was a serious danger, because HIV preferentially infects and kills immune cells. Even a simple skin infection can quickly turn deadly in those infected with HIV.

A neighbor was almost cheerfully paddling an inflatable dinghy up and down the flooded streets, rescuing people trapped by the rising waters. Like this local man, the people I met during my time in Thailand were extremely resourceful and independent, characteristics that were often coupled with a reverence for the concept they call *sanuk*, which loosely translates to "fun"—if it's not *sanuk*, it's simply not worth doing. It's also a coping mechanism for handling life's worst moments, like when your street and house are flooding. I learned a lot about *sanuk* firsthand that summer, and I saw how it helped people get through terrible circumstances, like caring for sick children.

The orphanage was a seventy-five-year-old teak structure that had recently been refurbished. It was surrounded by a lush garden with a pond, and the constant loud calls of birds made it easy to forget at times that we were living in the middle of a busy urban landscape. Housed within its walls were children who were the youngest victims of a growing epidemic that was starting to take a serious toll on the Thai population.

All the children at Tarn Nam Jai were born to mothers

who were HIV positive. At that time in the mid-1990s, many children were still becoming infected while in the womb. Transmission of HIV during an uncomplicated pregnancy was around 50 percent then, and that statistic was reflected in the children at the orphanage. (Recently, the Thai government has made immense strides in practically eliminating the mother-to-child transmission of HIV, and it's the first Asian country to do so.)

The idea behind the orphanage was to create a hospice for children who were HIV positive and an adoption center for children who were HIV negative. Since the test for HIV was still antibody-based back then, we had to wait at least six months to test the children to find out if they were infected. That's the length of time it normally takes for the mother's antibodies, which are proteins produced by specialized cells of the immune system called B cells, to clear out of a child's blood after they are born.

It was at Tarn Nam Jai that I really came to see how vulnerable young genetic boys were compared with girls. Anyone who has ever had to care for children is well aware that they get sick pretty often. This is, of course, much more extreme in HIV-positive children.

What was striking was that as infections swept through the home, it was often the HIV-positive boys who got sick earlier and much more seriously than the HIV-positive girls. Sometimes the boys got sick before the girls, even regardless of their HIV status.

I met Nuu and Yong-Yut early on during my time in

Thailand. Even though the two children were complete opposites, I would always find them running around and playing together. Nuu was quiet and cautious, which is why she was given her nickname, which means "mouse" in Thai. Yong-Yut on the other hand was always singing loudly, or finding new ways to annoy Nuu. His nickname means "strong fighter" in Thai, and it was given to him because he was always sicker than the other children.

It was obvious to me as well that Yong-Yut was much more susceptible to infections than his playmate was. That didn't make much sense, as they were both infected with the same human immunodeficiency virus. Every time a new microbe made its way through the orphanage, the more experienced staff started warning everyone, just as Rebecca would warn me in the NICU fifteen years later, to keep close watch on the boys.

I wondered at the time why the boys seemed so much weaker. It was only years later that I found out why Nuu seemed to handle her HIV infection better than some of the HIV-positive boys at the orphanage.

Today we know that even when they're treated with the same cocktail of antiviral medications called HAART,* HIV-positive women and men often have different outcomes. Antiviral medications like the ones in HAART interfere

---

*Highly active antiretroviral therapy (HAART) is a combination of medications that are used to treat people who are infected with HIV. Although HAART does not cure people who are infected with HIV, starting this therapy early typically allows for a longer life expectancy.

with the way in which viruses like HIV replicate, slowing down their growth and spread throughout the body. Since HIV preferentially infects and kills immune cells like CD4+ lymphocytes, reducing the number of circulating viruses allows the body's immune system to recover. Having a greater number of immune cells like CD4+ lymphocytes is important, as it allows us to fight off other opportunistic microbial infections.

Yet, just one year after starting HAART, significantly more men develop tuberculosis and pneumonia. Why is that? Just as with our faulty reasoning about men and ulcers, we used to think that some of the differences in HIV-infection outcomes and treatment were behavioral. Many thought that males were not responding as well as females to the HAART regimen simply because they were not taking their medication as diligently. But we now know that sex chromosomes play a role in how the body responds to HIV infections. For example, HIV-positive women have higher CD4+ lymphocyte counts than men in the early years of infection, which, as mentioned above, is an important marker of immunological strength. Women have also been found to have lower levels of HIV in their blood than men. This means that women's immune systems may be, at least initially, much stronger at fighting viral infections like HIV.

When the human body is invaded by a microbe, the backbone of the immunological response is our B cells' ability to make antibodies. B cells are essentially factories with the sole purpose of making antibodies that are specifically matched or fitted to a structure found on the invader called an immunogen. And the tighter the fit between an antibody and its immunogen, the better it works. Once B cells are activated and have seen combat, they retain a memory cell that can be called on years later if they're attacked again by the same microbe.

We employ this system every time we vaccinate someone. When we give patients an injection that contains microbial immunogens, we allow their bodies to make tight-fitting antibodies so that if they ever encounter that pathogen, they're one step ahead in the fight for survival. If we couldn't make a particular antibody tailored to a specific invader, we wouldn't survive for very long on this planet.

When a B cell produces an antibody that matches an invader, it "graduates" and moves on to refine its antibody for an even stronger and closer fit. The better the fit, the greater the chances are of surviving an infection. This antibody schooling usually takes place in the lymphoid tissue in the body.

Genetic females are uniquely evolved to make better-fitting antibodies that target microbial invaders. To make even better-fitting antibodies, B cells undergo a series of mutations. If the mutations occur in the genes that the body uses to make antibodies, they could end up making ones that have a better fit. When the B cells are being schooled to make better antibodies, mutations begin at a rate that quickly approaches one million times more than is normal, a process called somatic hypermutation. Both male and female B cells undergo this process of antibody refinement. Yet it is women who devote more energy to keep educating their B cells through more cycles of mutations until they are able to produce the best-fitting antibodies, ultimately fighting infections more effectively than men do. While there is more than one theory about why this hypermutation happens more effectively in women, one thing is clear: women have immunologically evolved to literally out-mutate men.

This could help explain why women are much better at making and using antibodies—their B cells are just more driven and capable of finding the best possible one. The X chromosome contains many genes that are involved in immune function. Women have two different X chromosomes in each of their immune cells that will contain different versions of the same immunological genes. So, women naturally have two populations of every type of immune cell, each predominantly using one X chromosome over the other. Having genetically diverse immune cells, like B cells, allows female cells to better compete to make the best antibodies possible. Men, of course, have nowhere near the same level of B cell competition, as they are all using the exact same X chromosome.

There's another reason that women might be more inclined to make better antibodies. Many women provide their babies with the antibodies they will need in the first months of life. The immune system of a fetus isn't fully activated while it is still in the womb, a likely evolutionary adaptation so that it doesn't start misguidedly attacking the mother. So, many mothers provide their babies with antibodies through breast milk, which delivers an immunological advantage. Studies show that breastfed babies even have a reduced risk for lower-respiratory infections years later as preschool children.

This whole process of causing more mutations to get a better-fitting antibody can also go horribly wrong—in males, that is. *H. pylori* can hijack the process of hypermutation and cause epithelial cells that line the stomach walls to unnecessarily mutate, which eventually can lead to gastric cancer. We still don't know exactly why, but men again seem to be particularly sensitive to this aberrant process.

IN APRIL 1924, just outside Vienna, a little-known writer was being tenderly attended to by his sister Ottla. The pangs of hunger that had become such a normal accompaniment to his waking hours were getting in the way of his work. But as his condition began to worsen, no matter how hungry he was, there was no way left for him to eat.

Like an Egyptian tomb in the process of being sealed, his esophagus was closing itself off to the world, and most crucially for Franz Kafka, to food. Kafka's digestive entombment was caused by millions of invisible microbes working their way through his laryngeal tissue. No wonder that the dreaded condition was named "consumption," as the victims often ended their once vibrant lives as unrecognizable, hollowed-out versions of their former selves.

Tuberculosis (TB) consumes its victims slowly over many

years. The disease has wreaked havoc on the lives of humans since the time we began domesticating animals. *Mycobacterium tuberculosis*, the infectious microbe that has killed millions of humans, is thought to have jumped species from infected cattle to humans some ten thousand years ago in the ancient Fertile Crescent—an area today spanning from Egypt to Iraq. But this isn't just some disease of the distant past: ten million people today are still infected with tuberculosis worldwide.

This wily microbe grinds down the body's defenses over time rather than fighting through an acute full-scale microbial attack. Once established, the lifelong chronic infection fights the body's immunological defenses through a process of attrition. Practically, this means that people who are physically weakened because of diabetes or from fighting another infection like HIV are much more susceptible to TB. This asymmetrical type of microbial battle is weighted in the direction of the attacker, and over time, it leaves those infected systemically withered.

The calling-card symptom of tuberculosis was a white handkerchief stained red from blood-tinged sputum. During the seventeenth through nineteenth centuries, about a quarter of all deaths were caused by TB. The industrial revolution in particular caused millions of people to start coughing up blood (the medical term for this is *hemoptysis*) from being infected with TB. Many factors contributed to the tuberculosis epidemic in the nineteenth century. Poor ventilation within housing helped spread the infection, a lack of proper nutrition suppressed people's immunity to the microbe, and even a lack of sunlight reduced the amount of vitamin D produced by the body.*

The classic symptoms of TB were described by Kafka, in a letter to his friend Max Brod, while it insidiously made itself at home in his body. He wrote: "Above all the fatigue increased. I lie for hours in the reclining chair in a twilight state . . . I am not doing well, even though the doctor maintains that the trouble in the lung has remitted by half. But I would say that it is far more than twice as bad. I never had such coughing, such shortness of breath, never such weakness."

As the TB kept spreading throughout his body, eventually invading his larynx, Kafka had to chew his food hundreds of times to be able to swallow without gagging. It's hard to imagine how uncomfortable the last few months of Kafka's life must have been.

Kafka was forty years old on June 3, 1924, when he eventually succumbed from the complications of tuberculosis. He asked Max Brod to promise not to read or disseminate his unfinished writings but instead to commit them all to the flames. Brod didn't listen.

As if coming across shards of broken pottery and piecing them back together, Brod assembled chapters and fragments into the completed works he imagined Kafka would have wanted. The truth is that we have no idea what *The Trial* and

---

*Emerging research indicates that vitamin D has an important role in supporting the immune system, which helps the body fight infections and malignancies.

his other novels would have been like if Kafka had lived to finish them.

What we do know with some certainty is that Kafka's odds of completing his novels were cut short because of the fact that he was a genetic male. Despite all of our progressive modern medical advancements, even as recently as 2017, almost two-thirds of the 1.3 million people who died from TB were men.

Another example of female immunological superiority in the face of disease is exemplified by an unfortunate incident known as the Lübeck disaster. In 1929, 251 newborns were accidentally given a dose of the Bacille Calmette-Guérin (BCG) anti-tuberculosis vaccine that was contaminated with *Mycobacterium tuberculosis*, the bacteria that causes TB. A significant number of the newborns who died after receiving the contaminated vaccine were boys.

Genetic females are really good at killing microbes.* One of the only bacteria that females seem more susceptible to is *Escherichia coli*. This is likely due to anatomical factors (not immunological ones), which make women more susceptible to urinary tract infections caused by microbes like *E. coli*. Fungal infections like those caused by *Candida albicans* are also more common in females for the very same reason. Given the anatomical differences between the external and internal genitalia of the genetic sexes, it's remarkable that females can fend off a host of so many invading microbes.

---

*There is a downside to genetic females having an immunological advantage over males, which I'll delve into more deeply in chapter 5.

Regardless of genetic sex, the biggest threat to our collective continued survival on this planet will always be infectious in nature. And even after the discovery of antibiotics more than seventy years ago, microbial pathogens still remain one of the biggest killers on earth. Just as TB maimed and killed in catastrophic numbers back in Kafka's time, emerging new strains of this microbe are now resistant to many of the antibiotics we have in our arsenal.

Multidrug-resistant tuberculosis (MDR-TB) is becoming increasingly difficult to treat for this very reason, as many of the antibiotics that once killed the microbe are now ineffective. An even greater threat is emerging from another strain called extensively drug-resistant tuberculosis (XDR-TB), which is immune to a greater number of antibiotics.

As people move around the world, so do microbes. XDR-TB has now been reported in 123 countries worldwide, including the United States. Knowing this is what led me to devote a large part of my professional career to developing new antibiotic agents to address the ever-growing threat of superbug, or multidrug-resistant microbial, infections.

When our immune system alone is not enough to deal with marauding microbes, we rely on antibiotics and antiviral medications to help clear infections. These medications are only a small part of the solution, because even when we employ the most advanced and powerful ones, all microbes eventually become resistant to them. As always, life manages to overcome any obstacles in its way. Which is why it's important to learn as much as we can about our existing innate system of immune defense. Even the best antibiotics and antiviral medications today don't "cure" our infections; instead, they provide only a little respite and assist us in the battle against microbes. It's still our very own immune systems that must finish the job.

Surviving in the pathogenic soup we live in is one of the biggest challenges we face as human beings. Be it triumphing over a severe bacterial infection, beating out the latest strain of influenza A, or more broadly enduring the trauma associated with famines and epidemics throughout history—women do it better. The reason has everything to do with their XX factor.

I say this as a geneticist and an antibiotic researcher: women truly are immune privileged. And that's a good thing, because our current and future survival on this planet depends on them.

3

DISADVANTAGED: THE MALE BRAIN

NAOMI WAS CLUTCHING A LARGE BROWN accordion-style filing case to her chest. Following closely and quietly behind her was her son, Noah. He was a rather tall teenager, striking in a demure sort of way. There was also a young woman around Noah's age seated in the waiting room. Noticing him, she looked up from her phone and stopped typing. But Noah wasn't paying attention to her, or to any of us. So it seemed.

"I have this recurring dream that it's early in the morning, and I'm sitting with Noah at our breakfast table," Naomi said

as she sat down in a chair across from me. This clinic's examination room was a little larger than most, and it even had
two windows. Although there was an empty chair right next
to Naomi where he could easily sit, Noah preferred to remain
standing behind his mom as she continued.

"I go on to chat with him about his favorite classes as well
as his after-school activities as he helps himself to another
bowl of cereal and then . . . he asks me if it's okay if he brings
his new girlfriend to our Thanksgiving dinner next week. And
just as I'm about to answer, I wake up . . . I've been having
different versions of this dream ever since Noah started high
school two years ago," Naomi said, her eyes beginning to water.

"The hardest thing for me about the dream is that I know
there's probably no way this will ever happen. I mean, Noah
stopped speaking when he was three years old and hasn't said
a word since. There's no pill, no diet or cure, that's worked
for him. I've long ago accepted Noah's condition. It's just too
painful for me to keep hoping for anything else. But every
night as I get into bed, I'm still left with a nagging feeling that
there might be something we missed. Maybe Noah's trying
to tell me something through this dream that I keep having.
Maybe if we take another look at his genes, I'll finally have
an answer?"

What Naomi brought along in her filing case that day
were copies of Noah's medical records from the time he was
a baby. There were detailed assessments from speech pathol-
ogists, psychologists, and his pediatrician. When Noah was
five years old, he was assessed by a specialist who determined

that he had autism spectrum disorder (ASD), which was assumed to be the reason he stopped speaking.

For a long time, the commonly held belief was that a boy was eight times more likely than a girl to be diagnosed with ASD. The reasoning behind this higher rate of ASD was thought to be that more boys were likely to get the medical attention that resulted in a diagnosis. For years this seemed to be true, as many of the cases of girls with ASD went undiagnosed. We just weren't aware that girls may symptomatically present differently than boys. Although that may account for some of the discrepancy in the diagnostic numbers between the genetic sexes, it definitely doesn't explain it away completely.

According to the 2018 numbers published by the Centers for Disease Control and Prevention (CDC), males are still three to four times more likely than females to be identified with ASD in the United States. We still don't know why the rates of ASD are so much higher in males, as we haven't yet fully explored the differences between the sexes. It may be the lack of another X in the brains of boys or their having a Y chromosome, or a combination of both.

When Noah was very young, he had some genetic testing done, which all came back normal, but Naomi already knew then that there wasn't a genetic test specifically for ASD. She explained to me that even if we couldn't do anything else for Noah, maybe there was a chance we could find a primary reason for his condition, now that genetic testing had improved.

After working through Noah's whole file, I ordered a comprehensive multigene panel and a few other tests and waited for his results to come in. They all came back negative, as they had previously, even with the expanded genetic testing that was now available. Naomi was understandably disappointed. We had learned nothing new about Noah's condition.

Although I stopped practicing medicine a few years ago, I still think about Noah, and so many other boys like him, and am struck by the myriad ways in which they've been challenged because of the hard fact that their brains have been shortchanged of an entire chromosome. The lack of X chromosomal variety within the cells that compose the brains of males makes them more fragile and sensitive to the insults of life, from infections to inflammation—conditions that are known to play a role in the development of intellectual disabilities. Whether or not this fact directly contributed to Noah's condition remains to be seen, but we know that when something goes wrong with an X chromosome, males don't have another one to rely on.

When all goes well, many of the genes on the X chromosome have the blueprints to make and maintain an optimally functioning brain. However, that doesn't always happen. Of the one thousand genes on the X chromosome, more than one hundred have been identified so far as a cause of intellectual disability. These conditions are clustered together under the term "X-linked intellectual disabilities." Many more genetic variations that reside on the X are thought to cause intellectual disability, but we haven't identified all of them yet.

Symptoms of X-linked intellectual disability present in early childhood and usually manifest with below-average intelligence. These conditions can be so severe that they prevent people from acquiring even the most basic skills necessary to lead independent lives. Other presentations of these conditions can be so mild that they're barely perceptible.

Geneticists know when the source of the intellectual disability originates with the X chromosome by examining the inheritance pattern in families of those who are affected. An X-linked inheritance pattern on a family tree stands out when only the boys in the family seem to be affected.*

When I was in the midst of working through Noah's file, I came across a letter from Noah's pediatrician summarizing some of the past medical investigations that she had undertaken. They included testing for a genetic condition called fragile X syndrome, which results in moderate to severe intellectual disability and affects males more often and more severely than it does females. Noah was suspected to have fragile X because his uncle (Naomi's brother) was affected with this condition. But Noah's testing for fragile X was negative.

Fragile X got its name because the X chromosome from these individuals appears to be more fragile and prone to breakage when observed under the microscope. Almost 99 percent of those affected with fragile X have an abnormality in their *fragile-X mental retardation* (*FMR1*) gene that stops it from working properly.

---

*Rarely, these conditions can affect females when they inherit two X chromosomes that have a mutation in the same gene.

The protein created from a working copy of the *FMR1* gene helps make connections between neurons (called synapses), which are crucial for normal brain development. Because individuals with fragile X lack the protein from the *FMR1* gene, their brains are improperly wired as a result. The medical consensus is that this wiring issue is the cause of many of the cognitive symptoms associated with fragile X.

Males are predominantly and more severely affected by fragile X because every single one of their cells, including neurons, is using the same fragile X chromosome—the only one they have. That's why the extra X chromosome that all genetic females inherit is crucial when it comes to protecting their brains. Not having the best genetic information to build and sustain one of the most complex biological systems that we know of can be problematic. As we've seen now over and over when it comes to the genetic lottery, it's always better to be able to play another hand.

We've known for a long time that boys are more susceptible to X-linked intellectual disabilities because they can't tolerate mutations on the X to the same degree that genetic females can. As I've mentioned, many of the one thousand genes found on the X chromosome are involved in the making and maintaining of the brain.

It's not just X-linked conditions and ASD that affect boys' brains disproportionately. From the start of life, boys are developmentally disadvantaged. The sex disadvantage for baby boys, which can result in lifelong neurological complications, was first reported by researchers all the way back in 1933 and still holds true today. We now know that boys are at a disadvantage when it comes to making the transition from the womb to the outside world, which results in a much greater risk of future developmental challenges.

Prematurity and distress at birth are both associated with an increased future risk of intellectual disabilities. An impressive study carried out in Finland tracked the health of every one of the 60,254 children who were born in the year 1987, until they all reached age seven. The study found that boys had a 20 percent higher risk of distress at birth, as well as an 11 percent greater chance of being born prematurely. The same study also found that as the children got older, boys had a two- to threefold higher risk for delayed development; and in a subset of more than 14,000 children, boys were at a higher risk for postponed school starts and the need for special education.

In a significant study published by the CDC in 2011, researchers looked at developmental disability data for children in the United States over a twelve-year period. The study found that "boys had twice the prevalence of any developmental disability and excess prevalence for attention deficit hyperactivity disorder, autism, learning disabilities, stuttering or stammering, and other developmental delays, specifically."

The latest numbers published for the United States by the National Center for Health Statistics likewise found that developmental disabilities are almost twice as prevalent in boys as they are in girls. This pronounced variance is not restricted to one geographic location or specific society. Many studies around the world are reporting findings similar to those from the United States, with an overall higher rate of developmental disabilities found in boys.

Even when you factor in the possibility of an overdiagnosis in males and under-identification in females in many of these conditions, males are *still* overrepresented diagnostically. Much of this has to do with the complexity involved in building and maintaining the human brain.

The brain is not a simple organ. And like everything else in the human body, it's built from instructions that are contained within the chromosomes and genes that we inherit. It's such a complex structure that even after it finishes its initial development phase, it continues to undergo remodeling—a process called neuroplasticity. These changes are ongoing until the day we die. Neuroplasticity is mediated not only by our DNA but also by everything we experience moment to moment. It's the reason we can still learn new skills long after we've left childhood behind.

There are many other structures in the body that, like the brain, are difficult to make, and it shouldn't be a surprise at this point that males, when compared with females, are also deficient in building these structures properly. These development mishaps can range from the somewhat benign to a more considerable congenital malformation.

When babies have trouble eating or sticking out their tongues, they may have tongue-tie, known medically as ankyloglossia. This is a condition in which the lingual frenulum, a piece of tissue that normally attaches to the underside of the tongue, is not properly connected. The tongue cannot move as freely as it should and is "tied" down in the mouth. Twice as many baby boys as baby girls are born with tongue-tie.

Clubfoot, or talipes—a condition in which the lower extremity doesn't form properly—is usually treated with physiotherapy and in extreme cases surgery. It's one of the most common birth defects in babies. Like so many other congenital anomalies, clubfoot is also twice as common in males, and we don't know why.

Almost everything that's biologically difficult to do in life, from survival to development, is done better by females. The genetic sex that overwhelmingly accomplishes the nearly impossible feat of becoming a supercentenarian is also the one with the least amount of developmental challenges. When you look around the world—country by country and regardless of culture—you see the same thing. Men's brains are at a disadvantage.

—

FEMALES NOT ONLY HAVE LOWER RATES of conditions such as X-linked intellectual disabilities, but they also have some superior abilities that are unique to possessing two X chromosomes. Some of these abilities are more obvious than we may realize.

I remember vividly how this played out once in my own life. My wife, Emma, and I were renovating our first apartment, and she came home one day excited to show me a handful of green Pantone paint-color palettes. Emma picked a few of the Pantone samples and placed them in front of me on the table. There was Parrot Green (Pantone 340), Crocodile Green (Pantone 341), and Leaf Green (Pantone 7725). They all looked pretty much identical to me. But she felt strongly about Leaf Green as the ideal color for the study. I had no idea what my wife was talking about. Could we really be seeing the world in different shades?

I know I'm not colorblind, but I *am* an XY male. Now, not all individual women have better color vision than men. It's just that they're not as likely to be visually color deficient and are much more likely to see greater color variation. At their genetic best, men can only aspire to have normal color vision.

Within the retinas of women's eyes are cells that are using only one of their two X chromosomes to build their receptors for color vision. That means some of the cells responsible for color vision might be using the X from their mothers, while others are using the X from their fathers. Having the use of two X chromosomes with different versions of the same genes explains why colorblindness is rare in females.

If the receptors making color vision that a woman inherits on each of her X chromosomes differ enough, it can result in a visual superpower. Scientists have yet to pin down the exact number of women who may see the world with enhanced color vision, but estimates range from 5 to 15 percent—maybe even more. This supercharged version of the color world is called tetrachromatic vision, and genetic females who have it can see one hundred million colors instead of the usual one million. A normal XY male has never had, and will never have, tetrachromatic vision.

We may not think of them in this way, but our eyes are actually outgrowths of our brain that grew into our face as we developed in the womb. And our eyes provide our brain with the information it needs to create a vision of the world around us. The incredible thing about our eyes is that they are not that fundamentally different in structure from the eyes of the earliest jawed fish that swam in the ocean some 430 million years ago. One of the major differences between what organisms can perceive in their environments is color. Light enters our eyes and hits our retina, but before it does, our cornea filters out most of the UV light. Our retinas are the screen on which the outside world is projected (inverted), and our brains then interpret that projected image. The cells that register the visible light are either rods or cones.

Rods absorb and respond to photons. The rods in our eyes (and there are about 120 million rods in each) register light. And within each rod cell, there are 150 million rhodopsin molecules that are packed into 1,000 discs.

We also have another 6 million cone cells in our retinas, and they work together to help the brain paint the world in vibrant colors. Most people have three different types of cone cells. Each type of cone cell uses a receptor from one of the three color vision genes—*OPN1SW*, *OPN1MW*, and *OPN1LW*—that helps it respond to a wavelength of light and relay that to the brain.

When one of these genes doesn't work as it should, it's more difficult for the brain to discern the difference between colors. If you lack a normally functional copy of one of the three genes that your retina uses to discern colors, such as *OPN1MW*, your ability to do so can drop dramatically, from around one million different shades to just ten thousand.

That's what happens in X-linked red-green colorblindness. And because two out of the three genes related to colorblindness are found on the X chromosome, males who don't inherit a working copy of these genes will likely "see" a much more subdued chromatic world.

There may be a small but significant upside to being colorblind, if we can learn something from the capuchin monkey. Scientists found that the colorblind male monkeys were much better at finding camouflaged surface-dwelling insects, which is great when you're looking for protein. This lines up pretty well with some anecdotal observations that colorblind human males are really good at "camouflage breaking," which means the ability to see through camouflage deception. As reported in an article in *Time* magazine from 1940, an army air corps observer could spot all the artillery pieces from midair that were camouflaged in a military exercise, while his peers struggled at the same task. How did he do this? Apparently, he was colorblind. This anomaly might come in handy in some situations, but when it comes to survival, seeing even more shades of color is priceless. And only women can do so.

Concetta Antico is a good illustration of the genetic superiority of women in this regard. Antico is not your run-of-the-mill visual artist. She has the unusual gift of seeing the world in millions of shades. In contrast to your average person, Antico sees about ninety-nine million more shades of color. She is a tetrachromat.

Most of us have trichromatic vision—the "tri" refers to the fact that we see the world using three separate genes that are used for color vision (two of which are found on the X chromosome). A tetrachromat like Antico uses different versions of the two genes for color vision found on her two X chromosomes.

Tetrachromatic vision exemplifies the power of cooperation that all genetic females have over genetic males. Even if not all women are full-blown tetrachromats, there's still a good chance that women have better color vision overall than your average man has.

Vision is so complex that it requires different cell types to cooperate to make it happen. It's not just the fact of having another X to choose from that allows a woman to see more colors than a man could ever dream of, but also the cooperation of the cells in her retina. It can allow females to do and see things that men can't.

Here is another way that genetic cooperation plays out in the world of vision: Long before the advent of farmers bringing produce to market, we had to put in an immense amount of effort to acquire fresh fruits and vegetables every single day. Have you ever wondered why your animal companion doesn't seem to have the same fresh produce requirements that humans do? It's because they can produce L-ascorbic acid, or vitamin C, independently and on demand. That's partly why they can survive on food made from such low-quality material.

It's not just cats and dogs that can do this (interesting aside: they are colorblind). Every other mammal on the planet, except for our primate cousins (and for some unknown reason, bats, guinea pigs, and capybaras), can convert the simple sugar glucose from their diet into usable vitamin C. So, what are primates like us to do? If we relied on our broken vitamin C–producing pseudogene* *GULOP* to do the job, we'd still have the same problem, because all of us have inherited broken copies. If you're trying hard to keep your teeth and stave off depression, inflammation, fatigue, and a host of other symptoms, as a human you will need to find some fresh fruit.

A visual system that allows its users to find fruit and even guess how ripe it is from a distance, without even tasting it, can be essential to survival for humans who lack the ability to make their own vitamin C.

But plants aren't about to give us a free handout. Plants can't easily locomote, so the evolutionary bargain they've worked out is that animals (including humans) get to eat their ripe fruit in exchange for "depositing" and planting the seeds *for* them. Fruit is expensive for plants to produce, so in exchange, they get a long and safe ride for their offspring.

To get our phytonutrient boost, including a good amount of vitamin C, we need to find fruit that's ripe. The plant's

---

Pseudogene is the term given for a sequence of DNA within the genome that resembles a working gene in related organisms but that has now lost its function.

way to signal this is often a color change in the fruit. If the seeds aren't ready and we eat the fruit anyway, then all the energy that went into making fruit is wasted. That's why fruit typically goes from a green that blends into the color of the background foliage to a more striking red, yellow, orange, or even a dark black—because then we'll be able to see and eat it.

Research on the behavior of one of our primate relatives, the wild trichromatic capuchins, suggests that they locate and eat fruit more quickly than colorblind ones do. Other research on captive rhesus macaques has found that the trichromatic females are faster than their colorblind peers at finding fruit.

If you're colorblind, it may be harder for you to discern when a fruit is ripe and safe to eat. In case you make a mistake, the plant has a smart and often somewhat toxic way of teaching you what's ripe, and that's taste. If you've ever bitten into an unripe banana, you'll know exactly what I mean.

WE DON'T USUALLY ASSOCIATE Japan with apples. Nor do we tend to associate the formation and maintenance of the human brain with apples either. My thinking around the similarities between neurological development and the pruning of apple trees in Japan has developed over the years as my work in the fields of neurogenetics and botanical sciences has deepened. In nature we often see examples of the same processes occurring at the macro and micro levels. I see the labor-intensive Japanese method of apple cultivation and the human brain in that way.

It was the apple harvest that brought me to Aomori prefecture in the middle of a brisk October. Aomori, located just south of the island of Hokkaido, is world famous for its apples. And few of the almost one million tons of the fruit produced there every year ever leave Japan.

Straining to reach up from underneath the thick canopy of a very large tree, I managed to pick my first apple. I was in Japan to collect tissue samples for a research project that was trying to tease out the genetic secrets behind a particular cultivar of apple. My project also focused on pruning efforts and how they change the way in which the genes in the apple tree behave. It was the ideal time for harvesting some of the most delicious apples on earth at their peak of ripeness. And it was impossible not to take a bite into some of my research subjects.

These red and juicy apples, of the Sekai Ichi variety, were definitely the largest ones I had ever seen. Their size is often matched by their weight, and the apples that day were no exception. Some of the ones we collected tipped the scales at just over two pounds each (for the sake of comparison, your average Red Delicious found on school lunch trays across the United States is just one-third of a pound). The size of these apples is not due to their genetics alone—a ton of human effort goes into making sure each Sekai Ichi apple gets this massive.

Joining me under the canopy that day was Naoki Yamazaki, a second-generation apple farmer, dressed for the part in a denim shirt underneath his blue jean overalls. His family has worked this same plot of land and tended to its trees for ages.

Yamazaki grew up eating pounds of apples every week and told me that he believes that he's literally made of apples. I asked him what he likes most about his farm. Stretching his arms out wide in both directions, he said, "My children," referring to the giant red apples hanging from his trees. I then asked him what was most challenging for him as a farmer. "Letting them go," he responded.

I've traveled and worked with many different farmers during the course of my research as I look for new biological compounds from plants and animals that we can use as therapeutics for human medicine. The one attribute that seems to unite them all is a love of their charges. It doesn't seem to matter what the farmers produce or grow. Whether you're an oolong tea farmer tending to wild and ancient trees in the Fujian province of China or a snail farmer from the Åland islands off the western coast of Finland, the sentiment is the same.

As I discovered on my visit to Aomori, Japanese apple-tree pruning is one of the more labor-intensive methods of farming, illustrating the cycle of life and death that occupies so much of a farmer's time. Yamazaki told me of a Japanese belief that it's only after pruning one thousand apple trees that you can call yourself a true apple farmer. When I asked him how many trees he's pruned, he answered, "Not that many."

The Japanese process of removing fruits was especially hard for me to watch because I love apples so much—it struck me as wasteful to pick and discard so many apples before they reached maturity. But this seeming wastefulness, believe it or not, changed my own thinking about the neurological processes in the brain. I now understand that pruning is a critical component of growing enormous, delicious apples, just as it's necessary in the cultivation of a healthy human brain.

During the process of pruning that happens throughout the year, a farmer will remove and discard branches, flowers, and immature apples that may be misshapen or bruised. This is done by hand. Yamazaki and his crew of workers pick through and discard hundreds of young apples as they methodically move through their orchard.

This allows each tree to focus all its attention on nurturing the remaining fruit. Yamazaki told me that it enables the remaining apples to grow to an ever-impressive size, filled with much more flavor. As they grow, all the apples are tenderly turned by hand, allowing them to develop a uniform red-striped finish. Although his overall yield of apples would be much higher without all the pruning, he said to me, "It's worth it. Sometimes less is more, no?"

Yamazaki goes above and beyond your typical pruning—he even gives each of his apples a little "shade cap" to wear for part of the season to make sure that they don't get too much sun. It's no wonder that these apples ultimately fetch about twenty dollars apiece.

Sitting and looking up at the branches from underneath the canopy of those apple trees in Japan reminded me of many of the parallel processes often found between nature and human development. Apples are not the only things in nature to benefit from proper pruning. A similar process needs to happen in the normal development of the human central nervous system, which includes the brain. Some neurons have to die so that others can survive and thrive.

For many years it remained a mystery as to how and why certain cells within the nervous system live or die. Then an intrepid and determined woman decided to shed some light on the whole process.

Dr. Rita Levi-Montalcini was out of a job. Italy had just entered the war on the side of the Axis powers in June 1940. The world around her by all accounts was alight with persecutory fires and had gone mad. Levi-Montalcini decided not to flee Italy, wanting to stay close to her family. As a Jewish woman, she found her future even more constrained. Her prospects of continuing to conduct neuroscientific research or work as a physician had looked grim since the latest "Laws for the Defense of Race" were passed on November 17, 1938, banning such activity. Hoping to help shield her from many of the restrictive laws against Jews that were being implemented, a friend from her medical school days even offered to marry her. She politely declined. Dr. Levi-Montalcini found herself spending all of her time thinking about the life and death of the neurons she was studying, and of her own very precarious chances of survival.

To keep herself busy, she started secretly working as a physician in Turin, a city in northern Italy where she was living with her family at the time. Eventually she had to stop practicing medicine altogether, as it was simply too risky. It wasn't just the war itself that she found distressing. Not being able to pursue her scientific questions deflated her sense of purpose. What she did next would come to typify Levi-Montalcini's life and work. Whatever the impediment before her, she always found her own way through, on her own terms.

Levi-Montalcini hadn't initially planned on becoming a physician or a scientist. Nor did she imagine devoting herself solely to raising a family like so many of her contemporaries. That women were not as capable as men in the sciences was the overwhelming dogma of her era. She wondered whether she was destined to be an artist like her twin sister, Paola. But for Levi-Montalcini, her scientific curiosity fueled the powerful creative engine that drove her forward.

Her main purpose in life, she would later convey, was born out of heartache. The woman who raised her, Giovanna Bruttata, was essentially a second mother to the Levi children. When Giovanna was diagnosed with terminal stomach cancer, Levi-Montalcini was devastated. It was at that moment she made her momentous decision to become a doctor. There were more than a few obstacles ahead of her. She was already three years out of high school, a time when most women her age in Italy were encouraged to find a man to marry and have children. In addition, her previous studies did not even begin to cover the fluency in mathematics, basic sciences, and ancient languages such as Greek and Latin that were a strict requirement to gain admission to medical school.

Levi-Montalcini prepared herself for the upcoming admission exam to gain access to what was predominantly a male domain. Many of her days started as early as four in the morning, with regular tutoring provided by local professors who were impressed by her intense ability to concentrate on the challenging new material. Eight months of long days of studying followed. But Levi-Montalcini was never satisfied with just purely memorizing information, and so even while still a student, the seeds of her later scientific questions were already beginning to germinate. The day of the admissions exam finally arrived.

Rita Levi-Montalcini earned the highest score among all those who sat to take the exam that day.

It's easy to see why someone with that level of tenacity would not let a few small things like a world war, or the dictatorial leader of her own country cozying up with the Nazi state hell-bent on the final solution, get in the way of her scientific research.

Harking back to the embryological studies of Aristotle and his work with chicken eggs, Levi-Montalcini doggedly pursued her research. Her plan was to study the embryological development of the human nervous system using fertilized chicken eggs as a model. She would dissect these eggs under the microscope and examine the chick's embryological development over time.

I've done similar laboratory work in the past, dissecting the respiratory system of sick honey bees, for example. It involved spending long hours hunched over a dissecting microscope, identifying and counting the tiny parasitic tracheal mites named *Acarapis woodi* that were living inside the bees. When these tracheal mites get into the breathing tubes, they make the bees' lives quite unpleasant. Imagine a fistful of small head lice making their way into your nose and then crawling down your trachea and into your lungs—and I'm sure you get the picture.

Honey bees need a lot of oxygen, given their level of activity, and they "breathe" by pumping the outside air through holes on the sides of their bodies called spiracles. Air makes its way through a system of tubes (which look like a collection of Slinkys of differing sizes) to the tissues and muscles that need it most.

My research involved unwinding the Slinky-like tracheal tubes under the microscope by teasing them apart using very tiny forceps and then counting each of the mites as they fell out. Progress was painstaking and slow. Spending hours partially frozen in an unnatural posture day after day not only was physically taxing but also required an immense amount of mental and physical perseverance. And my research project lasted only a few months. Levi-Montalcini's lasted a lifetime.

My work also took place in a pristine, brand-new, cutting-edge modern laboratory, with all the accompanying ergonomic bells and whistles.

Now contrast that with Levi-Montalcini working away in the hardscrabble laboratory that she built for herself during the war. Her private laboratory, which she named "à la Robinson Crusoe," was put together with the help of some friends in her small bedroom in her family's flat.

The microscissors she used to cut tissue under the microscope came from an eye doctor, and the minuscule tweezers were from a watchmaker. Miniature scalpels were apparently much harder to come by, so she had to make them by hand. Levi-Montalcini used these tools together with a microscope to make dissections and stained slides to document what she was seeing. She also put together a makeshift incubator to keep her eggs warm as she watched them develop. Finding fertilized chicken eggs wasn't always easy either.

In vertebrates like humans, nerves exit the spinal cord and innervate our limbs. This allows our brain to stay informed about what our limbs might be up to, through all kinds of sensory feedback such as temperature and vibration. We also move our limbs through the nerves that innervate our muscles, such as the ones in my hands and fingers that I'm using right now to type these words.

If these nerve cells aren't wired correctly to our muscles or skin as we develop, or if they are severed in an accident later in life, we can lose the sensation in our bodies and lose control of our movements. What Levi-Montalcini discovered was that there had to be a chemical agent, a secret chemical key, that was keeping the nerves alive as they grew into the limbs she was dissecting. Today we call that mysterious and previously unknown protein compound nerve growth factor (NGF).

We now know that there are many other types of proteins involved in the regulation, development, function, and survival of neurons. Collectively we refer to these proteins as neurotrophins. Other important neurotrophins that have been identified include brain-derived neurotrophic factor (BDNF), neurotrophin-3 (NT-3), and neurotrophin-4/5 (NT-4/5). Many of these neurotrophins are also thought to be involved in a myriad of neurological conditions, including everything from Alzheimer's disease to autism spectrum disorder and even attention deficit hyperactivity disorder. Many recent studies on nonhuman animals examining the function of neurotrophins have found them to be sex dependent. This is relevant because neurotrophins like BDNF are not small biological players. They have significant effects on every aspect of the way the brain works, from neuronal survival, which was discovered by Levi-Montalcini, to dendritic branching and synapse formation, which we'll look at in a minute. Many of these neurotrophins also interact with and are driven by inflammatory processes.

The amount of neurotrophins present within our bodies can also be influenced by our lifestyle choices. Whenever we moderately or intensively exercise, neurotrophins like BDNF become elevated, which helps to keep the brain optimally functional. We're only beginning to understand how many of these neurotrophins work, and what we do know is largely due to the pioneering work of Levi-Montalcini.

After the war ended, Levi-Montalcini joined forces with the biochemist Stanley Cohen, who eventually worked out the structure of the mysterious compound NGF she had originally discovered. With his dog Smog—according to Levi-Montalcini, "the sweetest and most mongrel dog I ever saw"—often by his side, Cohen would pay many visits to her laboratory. Learning from each other, as Cohen knew little of the nervous system and Levi-Montalcini wasn't as familiar with biochemistry, their work together proved fruitful.

More than forty years later, for the work that started during the middle of a world at war, Rita Levi-Montalcini was awarded the Nobel Prize in Physiology or Medicine in 1986, along with Stanley Cohen. This major breakthrough in scientific understanding set the stage for more scientists to grasp the life-and-death cycle of neurons, as well as to begin to appreciate some of the most basic differences between the sexes.

WE ALL START OUT with billions of neurons, and even more synaptic connections between each of them. Just as Yamazaki's goal was to get better apples through pruning, for normal human brain development to occur, a lot of very careful cellular and synaptic pruning has to take place. This is why, overall, we end up with fewer neurons in our brains later on in life as adults than what we started out with as babies.

The human brain is massive and metabolically expensive to run, and it greedily consumes around 20 percent of the calories we burn every day just to stay functional. For most of our evolutionary history, a regular meal was never guaranteed, so having a big brain that needed to be fed even when food was scarce could be problematic.

Evidence from various neuroscientific studies has demonstrated that the human neuronal pruning process begins very early in our development—while we are still in the womb, in fact. This biological technique—an overproduction of neurons and then a broad pruning of them—is a process that has proved successful. Imagine a kitchen drawer overfilled with utensils. Having more utensils isn't always helpful—they may in fact just make finding what you're looking for all the harder. The goal of normal brain development is akin to the minimalistic dictum known as "use it or lose it," and this goal makes it easier for better neuronal communication to take place.

Having neurons around that are not contributing to survival and are an energetic drain is not helpful. A solution to make the brain work more efficiently is to prune or kill those neurons. Connections between neurons that are not being used very frequently can be cut back as well. That's the unwritten law of biology—Mother Nature runs a tight ship.

Some of the latest neuroscience research is implicating a specialized type of immune cell that lives within the brain, called microglia, in a host of neurological conditions. We used to think that microglia's only role was immunological—that their sole purpose was fighting microbial invaders and similar threats. Microglia cells will try to take out anything that's foreign—bacterial or viral. Microglia are plentiful within the brain. Almost 10 percent of the billions of cells of the nervous system are microglia.

We now know that microglia are working in the brain—much like the apple farmers are working in Japan—not only fighting against infections but also making their way through the thicket of neurons, snipping away and removing the underused connections between them.

This microglial synaptic-pruning process is a recent discovery, and it has changed the way we understand the normal development of the brain. This theory had been around the edges of the scientific community for some time, but it wasn't until March 2018 that researchers finally managed to capture images of microglia cells actually doing this critical pruning work. As with my own experience witnessing the pruning practice of Japanese apple farmers, brain researchers saw firsthand that the microglia were inducing structural changes and re-arrangements of synapses.

We now think that microglia play a role in the neurological and autoimmune disorder multiple sclerosis (MS). Microglia are more active when there's inflammation, as is the case with MS. Like almost all autoimmune conditions, MS predominantly affects females over males, which, as we'll discuss later, is one of the drawbacks of having a better immune system. Microglia are derived from the immune system, and we've established that immune cells behave differently in women and men. What we don't know at this point is how microglia might be behaving differently *inside* the brains of men and women.

Now that they've piqued our interests from a disease perspective, misbehaving microglia are being studied in everything from traumatic brain injury (TBI) to the development of Alzheimer's disease and even autism spectrum disorder (ASD).

A relatively large and recent postmortem study looked at the brains of individuals with ASD and found signs of chronic inflammation, which was thought to have been triggered by microglia. What was even more interesting was that certain areas of the brain that have already been flagged by neuroscientists as relevant in ASD, such as the dorsolateral prefrontal cortex, were particularly affected by an inappropriate microglial response. This is important, since the dorsolateral prefrontal cortex is involved in executive function (higher-level decision making)—something that's not always working in the same way in people with ASD.

What we don't know for certain is what's driving microglia to be unruly and ultimately what role microglia play in the development of ASD. In other words, are the microglia pruners acting out on their own or under instruction provided by another biological process? And what's not yet clear from all this emerging research is what microglia are up to when life's biological processes are proceeding normally. Are they just pruning and maintaining synapses, or are they also providing the tender microscopic support I saw in a similar macroscopic vein on Yamazaki's apple farm?

So much more is waiting to be discovered. But what we do know for certain is that being born a genetic male with only one X chromosome significantly increases the chances of being diagnosed with ASD.

—

THE FIELD OF GENETICS today is still a bit like my toddler nephew. Now that he's successfully acquired a few words, he's in the process of combining them to make short but meaningful sentences about the world around him. Just like my nephew's understanding of English, geneticists understand basic genetic words and "commands" but are only just beginning to grasp the subtle intimations of genetic knowledge—to say nothing about how all that knowledge can be translated into clinical practice.

That's the reason genetics is so ripe for interpretation. As with boomtowns that spring up during a gold rush, entire industries have appeared seemingly overnight to try to help us make some sense of our genes. There's been much promise but little delivery so far when it comes to the ever-burgeoning business of commercial genetic testing. Millions of people around the world have already opted to mail in samples of their DNA in the hope of having their ancestry revealed to them. What most of us don't realize, though, is that our ancestry results are more dependent on the algorithm that the company uses for its analysis than on our actual genetic ancestry. Some companies even go so far as to customize exercise regimens or promise to help you find that perfect mate—all based on your genes. The reading of genetic tea leaves is big business.

But our DNA just keeps on doing what it has done for millions of years. Our genes alone do not totally dictate our lives but are constantly reacting to the world around them and responding in turn. Imagine two identical Steinway grand pianos on a stage with the same piece of sheet music for Beethoven's *Moonlight* Sonata resting on both. Two musicians approach each of the pianos separately and begin to play. Musical notations as well as instructions on how it should be played dictate what we hear when we listen to a piece of music. Although each musician is playing the same sonata, the way in which the two are playing the musical piece can sound vastly different.

Far from being schematic with clear instructions, the human genome is written in a way that we are still trying to fully comprehend. We know that about three billion nucleotides—noted as adenine (A), cytosine (C), guanine (G), thymine (T)—are strung together like pearls on a DNA necklace that encodes for genes that are essential for life and for others that are more mundane. All the information—from whether or not someone needs to use deodorant daily (which is encoded in the *ABCC11* gene) or whether cilantro tastes like soap or is just plain delicious (which is encoded by variants of the *OR6A2* gene)—can be found within the human genome.

We are constantly using a repertoire of genes within our genome to meet the requirements of diverse situations. And cells within our bodies use some genes more than others, depending upon what's required at the time. Responding as needed genetically to the constant changes and challenges of life has allowed us to survive for this long as a species. Having two X chromosomes instead of one allows women to have more genetic instructions with which to react more creatively to life.

I THOUGHT A LOT about the choices we make in life and how our genes respond in turn to these changes after I met Paul. His story exemplified for me how the sex chromosomes we inherit can shape the parameters of the choices available to us.

Back in the 1960s, there was a lot of emphasis on the Y chromosome as the reason for men's misbehaving. The thinking behind much of the research associating violence with having a Y chromosome was not altogether wrong.

Elevated levels of androgens like testosterone (which men possess as a result of having a Y chromosome) definitely play a role. But men's disadvantage may not be just the burden associated with inheriting a Y chromosome itself. Beyond having higher levels of androgens, males simply don't have the same genetic options available to them as females.

By all accounts Paul was very successful professionally. When we first met, he was in his midfifties and had already managed to make a small fortune deftly investing his clients' money. Paul had even seen his clients through the most recent global financial crises, emerging relatively unscathed.

The investment firm he had started only a few years out of business school with two of his closest friends was now reluctantly turning away prospective clients because it was just too busy. Happily married with two teenage children, Paul devoted some of his free time to philanthropic organizations that needed financial guidance.

I was flying back to New York after a long research trip abroad. It was always nice to be back home after several weeks away. I turned on my phone after my plane landed at JFK and saw that I already had two urgent messages from Paul's office. His assistant asked if I was available to meet with him for an early breakfast the next morning. Paul wanted to speak with me about some confidential genetic testing he'd had done.

I wasn't able to see him right away, but we got something on the books without delay. Over breakfast a few days later, Paul handed me a thick manila envelope. As I thumbed through the stack of pages, it became clear why Paul was asking about his genetic testing.

What I was looking at were the results of anonymous genetic testing that Paul paid to have done through a genetics research laboratory. The section highlighted in yellow was likely the reason that he wanted my opinion.

While I read through Paul's paperwork, I looked up at him and noticed that he looked pensive.

"So what do you think, Doc?" Paul asked. He didn't waste any time. What his results indicated was that he had inherited a rare version of a gene called *MAOA*. The reason he was seeking my expertise was that the genetic change he was found to have inherited in his *MAOA* gene was marked as a variant of unknown significance (VUS). This is geneticist-speak for a testing result that may be something—or absolutely nothing.

Variants of unknown significance are a prime example of why we are so far from understanding all the implications that result from the genes we inherit. For Paul, this VUS resulted from a change in the *MAOA* gene that no one had ever seen before, and because of this, no one at the company who was reviewing the results was sure what to make of it.

What we do know about the *MAOA* gene is that it encodes for an enzyme by the name monoamine oxidase A. This enzyme chews up and recycles neurotransmitters like serotonin and, to a lesser extent, noradrenaline and dopamine. As with most genes in the human genome, when the *MAOA* gene is working, you don't notice much. When the *MAOA* gene isn't working, however, things can quickly spin out of control and without much forewarning.

This is exactly what a Dutch geneticist by the name of Dr. Han G. Brunner discovered and reported in 1993. Brunner was very interested in a family with multiple male members who committed extreme violent acts and exhibited impulsive aggression. No one could explain exactly why all these men behaved so violently. They also seemed to have some level of cognitive impairment and intellectual disability.

What Brunner found was that all the men had the same mutation in their *MAOA* gene. This point mutation was a single nucleotide, or "letter," change in our three-billion-letter-long genome code. This was enough of a difference to cause a complete deficiency of the *MAOA* gene product, and therefore all the differences in behavior that Brunner witnessed. Since Brunner's original report, researchers have genetically engineered mice without the *MAOA* gene and found them to be more aggressive as a result. As with the patients Brunner first described, the mice in this study provided scientists with more evidence to confirm the important influence that this gene can have on behavior.

In most people, the *MAOA* gene is functional. There are two common versions of this gene that most of us inherit: a low-activity version and a high-activity version (written as *MAOA-L* and *MAOA-H*, respectively). Of the two, the high-activity version is much more common. The low-activity version of the *MAOA* gene is thought to result in a slow-acting *MAOA* gene product, meaning that neurotransmitters like serotonin are not recycled as quickly as they would be with the high-activity version.

Mischaracterized as the "warrior gene" in the 1990s, the *MAOA-L*, or low-activity version, is not without controversy in the field of genetics and in society at large. Some scientists think that it makes a person more prone to antisocial behavior and criminal violence. Studies in humans have found an association between the low-activity *MAOA-L* version of the gene and greater and more frequent acts of aggression, especially in people who have experienced forms of maltreatment such as child abuse earlier in life.

Because the *MAOA* gene is found on the X chromosome, genetic males inherit only one copy, while females inherit two. This means that males who've inherited the *MAOA-L* version may be responding in a more extreme way to negative experiences than females do, since they don't have another copy that is modulating its effects.

Paul started reading about the *MAOA* gene and came across an article that referenced Brunner's patients. Not making any progress when he sought more information directly from the research laboratory that did his genetic testing, he was left on his own. He also read an article that described the *MAOA* gene as the "psycho gene," so it's understandable that Paul was becoming concerned about his genetics.

Although his behavior was nowhere near as extreme as that of the patients Brunner described, Paul did share with me that he had always struggled with what he called his "explosive rage."

Somehow, in the past he had always found a way to manage what he called the "Hulk side" of his personality. "My wife's been my biggest supporter in this regard. Looking back, I'm scared to even imagine what I would have done without her in a few situations that could have quickly spun out of control," he said to me. He did have a history of uncontrolled outbursts, especially after he felt slighted in some way. This had only gotten worse with time. He told me that his anger issues were amplified recently when his business partners said they wanted to expand. Paul wondered whether the trouble he was having reining in his temper had anything to do with his genes.

He wanted to know if I thought he might have inherited Brunner's syndrome. I thought for a moment and responded,

"I don't think so . . . You've had significant accomplishments both professionally and academically, and those who are affected by Brunner's syndrome often exhibit some type of cognitive impairment and episodes of extreme behavior."

"Does this result mean that I've inherited a problem with my *MAOA* gene? Like it's not working properly?" he asked.

I listened to Paul's concerns and recommended he have his genetic results run through some predictive algorithms. I didn't see any of that information in his current genetic results, and I thought it could help. I also suggested that he look into whether his VUS results had been reported in the scientific literature recently. If it had been reported since he'd had his testing done, the findings would help him better understand the significance of his VUS results. There were also some additional laboratory tests that I felt he could undergo.

I explained to Paul that this would involve having his doctor order biochemical tests to measure the serotonin in his blood and also the breakdown products of neurotransmitters that might serve as a proxy for the functioning of his *MAOA* gene. I also referred him to a research paper, in which individuals with a low-functioning *MAOA* gene reported an improvement in their symptoms when they were prescribed a selective serotonin reuptake inhibitor (SSRI) drug such as fluoxetine or Prozac. When the *MAOA* gene isn't working well, the result is higher concentrations of neurotransmitters like serotonin in the synapses between neurons. So, you would think that knowing this, the last thing you would want to do is to give people in Paul's situation an SSRI drug that increases the amount of available serotonin even more. Paradoxically, taking an SSRI has been reported to actually improve symptoms.

And then I said, "I have to be honest with you, Paul. The truth is, you may never know exactly how your *MAOA* gene is behaving in you."

"And what about my two daughters? Do I need to worry about them?" he asked.

"Probably not. Since the *MAOA* gene is on the X chromosome and since they have more than one copy, they're likely to be protected even if the gene they inherited from you isn't working. That's why in the original family that Brunner described, it was only the males that were aggressive even though some of the females inherited the same copy of the mutated *MAOA* gene."

"Like with colorblindness? I'm colorblind, but my daughters are not."

"Right," I answered. "Your daughters are protected by having another copy of the X chromosome, and the cells within their brains can rely on the X with a working copy of the *MAOA* gene. Paul, as a male, you just don't have those options."

I could see clearly that Paul was hoping for more certainty. I tried outlining the best way to proceed in case he wanted to pursue the matter further. A few months later, I got a call from Paul's assistant. Paul had asked her to let me know that he was doing much better. There was no real way for me to know whether or not the SSRI that his doctor prescribed for him was helping because of the version of the

MAOA gene he inherited. Either way, I was very glad to hear his update.

When it comes to the immense complexity of human behavior, we still don't know how exactly genes come together and work with the environment to mete out their effects. We're much better at understanding or predicting what happens when things go wrong, as in the case of Brunner's research showcasing males' violent and impulsive behavior.

During the Vietnam War, the Buddhist monk Viet Tong experienced considerable trauma. He also happened to be in possession of the low-activity version of the *MAOA* gene. Something he said summed up, in my mind, the entirety of the field of behavioral genetics: "Everyone is born with good and bad traits, that's what makes us human. But everything in life is not set in stone, our future is constantly changing, it's what we do now that will affect our future tomorrow."

Often, men will have to work harder to deal with the hand that their genetics has dealt them. That's going to be especially true for the hundreds of genes that are found on the X chromosome, as many of them are involved in the way the brain forms and functions. And this is exactly what Brunner saw with the original family he described.

Paul's case reminded me how women, unlike men, already possess an ability to modulate their behavior in a genetically built-in fashion. All genetic males have brains that are using the exact same X chromosome, which is why so many more males have X-linked intellectual disabilities. Genetic females, on the other hand, have brains that are using the genetic information provided by two X chromosomes. That means if a male inherits a version of a gene on his X chromosome that affects his behavior—such as the one that causes Brunner syndrome—he will always be affected. Females are rarely, if ever, affected by Brunner syndrome. The ability to use more than one X chromosome in their brains helps females dampen any ill effects of having a mutation on either one of their X chromosomes.

Females have genetic options. I believe this to be the underlying mechanism that explains why genetic males are overrepresented in a host of conditions like autism spectrum disorder, intellectual disability, and countless other developmental delays. And because genetic females have two different X chromosomes active in their brains at all times, they can better deal with the consequences of mutations in genes like *MAOA*.

It is this lack of both genetic options and cellular cooperation that creates the conditions for an XY male to suffer from so many types of disadvantages, like the ones I've outlined in this chapter. Women don't experience these conditions to the same degree as men because of their robust genetic endowment, which positions them to make superior genetic choices.

4

STAMINA: HOW WOMEN OUTLAST MEN

T HE TERRACES OF BAYCREST in northern Toronto, Canada, is a bustling community of active and animated older adults. The environment certainly doesn't give you any notion that human life can be short. There's a constant state of healthy movement at the Baycrest facility, with a recreational schedule filled with engaging and challenging activities. Residents report that these activities not only are meaningful but also encourage them to cultivate skills that they may never have even considered possible to acquire at this stage of their lives.

None of our ancestors could have envisioned a future for themselves that collectively involved such relative health and longevity. But here at Baycrest, as at countless other centers catering to older adults worldwide, the numbers tell us a very different story.

Our life expectancies have been increasing significantly over time. Japan, for example, currently has one of the populations with the longest-lived individuals: current life expectancy is 84.2 years of age. Even Afghanistan, a country with one of the lowest life-expectancy levels, hovers around 62.7 years of age. That's still significantly longer than what a seventeenth-century Londoner might have experienced, when life expectancy was only about 35 years of age.*

With all our advancements and improvements when it comes to elder care, there is one significant thing that you would notice when walking into the Terraces of Baycrest. You may have guessed it. There aren't many men.

For most of our collective past, we thought that death treated both sexes equally. We were too distracted by the daily suffering that surrounded us to notice that the Grim Reaper was actually discriminating between the sexes. Hunger, pestilence, violence, and climatic upheavals have come and gone throughout human history, but women have always outlasted men. Scarcely a century ever passes without a great

---

*People did make it to their sixties and beyond in London at the time, but the chances of this for the average individual were slim, given the high rate of infant mortality.

calamity befalling humanity—either environmental, micro-bial, or both—and genetic females outlast males every time. It happens at the beginning of life, at the end of life, and throughout life as well.

In a book published in 1662 titled *Natural and Political Observations Made upon the Bills of Mortality*, the Englishman John Graunt provided the first statistical proof that women outlive men. As a hobbyist statistician and demographer, he studied municipal death records of London parishes.

And there was no shortage of death in Graunt's seventeenth-century London. During the Great Plague of London (1665–1666), around a quarter of the city's population is thought to have been felled by the plague. No one knew for certain why some years were worse than others. To help predict and track untimely deaths, a few brave elderly women called "searchers" were appointed and employed as a form of parish charity to inspect the recently deceased and determine their cause of death.

This was important work at the time, since it could alert the authorities to a brewing plague or other such calamity. Parish clerks sold this information to seventeenth-century Londoners who were more than eager to pay for the privilege of knowing if and when death might strike. The documents, known as the Bills of Mortality, could be purchased individually or at a discounted rate by weekly subscription. The bills comprised the weekly tally of all the burials and christenings, compiled from parish registers.

Some printers even published the number of dead from

prior plague years in London to make it clear that mortality had a seasonal pattern, with deaths peaking in the summer. This allowed the avid reader of the weekly bills to compare current numbers of the reported dead to past trends. At that time in London, death was good business. Much as we rely on quarterly finance reports today to make decisions regarding our financial portfolio, people back then knew that when the Grim Reaper was back, it was a worthwhile investment to escape the city in time to avoid death's imminent clutches.

Graunt also relied on the data from the Bills of Mortality and subsequently discovered that there was a considerable discrepancy in life expectancy when it came to men and women. At the time Graunt started picking his way through death records, people the world over had no idea that there might be a difference in longevity between the sexes. Why would there be when the average life expectancy for both sexes hovered around thirty-five years of age, and men were consistently favored over women in virtually every arena of life? Men were ultimately presumed to be the stronger and healthier of the two sexes.

The seventeenth-century Englishman Edmond Halley (who also successfully predicted the return of a comet that today bears his name) was, like Graunt, focused on longevity research and published his own findings in the *Philosophical Transactions of the Royal Society* in 1693. His life table was based on the 1687–1691 demographic statistics from the city of Breslau, which today is the Polish city of Wrocław. Unlike

Graunt, Halley pooled male and female survivorship data and found that overall survival decreases with age.

Halley's paper was a highly significant contribution to demographic statistics because it illustrated to those who sold life insurance that they had to take the age of the buyer into account. His work was ignored for decades, but with time, it was finally recognized that when it comes to making money, the use of discriminatory practices against the aged pays. Everyone selling life insurance eventually realized that they would risk bankruptcy if too many short-lived people bought their policies.

The significance of Graunt's findings regarding the survival of women over men was also eventually appropriated and used by those who made their living selling life insurance.* It wasn't just the age of a person that mattered, but that person's sex as well.

No matter where you look around the world, when it comes to longevity, women always come out on top. Japan sees women living to an average of 87.1 years of age, while the average Japanese man can expect to live to 81.1 years. In Afghanistan, men live on average for 61 years, while the average for women is 64.5 years. If we look at the most long-lived of humans, the supercentenarians—a group of people who have reached

---

*Knowing and accepting the differences between the sexes when it comes to longevity pays off well for companies selling life insurance, especially when premiums are determined by the likelihood of death. That is the case if insurance companies are operating outside the European Union. Discriminating by sex to determine the cost of premiums for life insurance is currently illegal only in the EU.

110 years of age—95 percent of them are women. The female survival advantage is clear.

———

THROUGHOUT HUMAN HISTORY there are many examples of the female survival advantage. The story of Marguerite de La Rocque is just one of them. Marguerite was twenty-six years old when she found herself on the journey of a lifetime. It was April 1542, and she was sailing from France to what we know of today as Canada, along with her relative Jean-François de La Rocque de Roberval, who was the captain of the voyage. With the ease of global travel today, it may be difficult to imagine the overwhelming excitement and trepidation Marguerite felt as she left behind her old life in France to start over in a whole new world.

Marguerite's trip from France was initially without incident—no small feat given all the dangers in undertaking such a journey. But things became complicated when Marguerite developed a romantic involvement with a fellow passenger. As punishment, Roberval abandoned Marguerite on a barren island off the coast of Canada. Leaving the French ship, Marguerite was accompanied by her lover and a loyal maidservant by the name of Damienne, as both refused to let her die alone. Roberval instructed that the group be given a musket and a few provisions. In this way, he was perhaps hoping to absolve himself from what amounted to a death sentence for them.

Speculation as to why Roberval would have behaved so cruelly to his relative and charge ranges from his desire to strictly adhere to his secret Huguenot Calvinist religious beliefs to financial greed. The latter is likely true, because as soon as Roberval made his way back to France, he maintained that Marguerite was dead and then proceeded to claim all her inherited wealth and property.

The three castaways were now alone on a small and rocky uninhabited spit of land situated at the entrance to the Gulf of St. Lawrence. The Island of Demons (today thought to be Belle Isle) was a fitting name for their new home, as it was next to impossible to find food or shelter. Very quickly, whatever provisions they were given ran out, and Marguerite, her lover, and maidservant began to starve. Not only that, but they were about to be joined by another mouth to feed, as Marguerite found herself pregnant.

Marguerite's lover and then her maidservant perished. Stranded on the island, Marguerite was alone for the first time in her life. Somehow, she managed to deliver her baby without any assistance and, miraculously, both mother and child survived. But not for long. Her milk supply dried up and her baby boy, just one month old, died. Yet that was not the end of Marguerite.

Three long years later, Marguerite was rescued by a group of passing Basque fishermen. We are told they found Marguerite draped in the skin of a bear that she had shot and killed. They couldn't understand how she survived alone for such a long time in such an inhospitable place.

Soon, providence caught up with Roberval. He was confronted by an angry French Catholic mob after leaving a secret Huguenot Calvinist meeting. Roberval was attacked, beaten, and killed. Marguerite, on the other hand, returned to France and opened a school for girls.

The story of the Donner Party is another equally striking example of the female survival advantage. While many people know the broad outline of this story, it gets even more interesting when we examine the details. Attempting to travel by covered wagon from Illinois to California close to the start of winter had some members of the Donner group expressing outward trepidation before they even departed. Their concerns seemed prescient when all eighty-seven members of the group were suddenly trapped in the Sierra Nevada wilderness by a blizzard that struck on October 1, 1846.

What's remarkable is that nearly twice as many men as women died: around 57 percent of the men, compared with 28 percent of the women. The men also succumbed much more quickly than the women. Given their circumstances, it's amazing that any members of the Donner Party survived at all. Running out of food and exposed to the elements, some of them, infamously, even resorted to cannibalism.

Marguerite's survival on a deserted island and the female survivors of the Donner Party journey, along with Graunt's demographic data, might seem like historical outliers. So let's move from looking at individual survivorship and early examples of demography to more recent patterns of survival within and between populations.

During the years that the Soviet Union had dominion over modern-day Ukraine, a policy of collectivization was instituted at the behest of Joseph Stalin. The stated aim of eliminating individual peasant farms and replacing them with collective ones was to increase food production and, in so doing, the availability of produce to urban workers.

What resulted was one of the worst man-made demographic catastrophes in recent human history. Instead of an increase in food production, a horrific famine ensued. Parts of modern Ukraine were especially hard-hit, with an estimated six million to eight million people perishing between 1932 and 1933. Among the suffering of millions of people, caused by a Soviet collectivization cataclysm, we now know that women were still outliving men. Life expectancy for Ukrainians prior to the calamity was around 45.9 years for females and 41.6 years for males. After the famine ensued, life expectancy nose-dived to a startling 10.9 years of age for females and 7.3 years for males. Notice something here?

Think again about the overwhelming majority of elder women you see anywhere in the world, and the survival advantage is undeniable.

We used to assume that the only reason behind the early demise of men was behavioral. We now know that this survival advantage begins shortly before birth. The female survival advantage between the sexes still holds regardless of education, economic factors, and alcohol, drug, or tobacco consumption. Genetic males might have more muscle mass and greater height, overall size, and physical strength compared with their genetic female counterparts, but when it comes to surviving physical hardships, women almost always outlast genetic males.

Yes, of course, there are still behavioral differences between the genetic sexes such as risk-taking behavior. But that is not the whole story. Data from the nineteenth and early twentieth centuries about Mormons living in Utah, who eschew both tobacco and alcohol, revealed that women affiliated with the church still outlived their male counterparts. This was the case even though at the time these women had an even higher average birthrate than that of the general population, which meant an increased risk of death with every pregnancy.

Men do outnumber women in the occupations that are the most dangerous. According to the U.S. Bureau of Labor Statistics, from 2011 through 2015, men accounted for 92.5 percent of all workplace deaths. Yet a study from Germany also found a female survival advantage within a data set size of over eleven thousand—a particularly large sample size—of cloistered Catholic nuns and monks. Using mortality data from 1890 to 1995, comparing Catholic nuns and monks in religious orders in Bavaria with the general German populace, the researchers still found a longevity gap that favored females. This was the case even though the monks were living in a relatively closed community and so were not exposed to the same types of labor and lifestyle risk factors that were experienced by the average German man. Obviously, the mechanisms driving this chasm between the chromosomal sexes when it comes to survival, development, and aging are deeper than just behavior.

━━

THE STORY OF HUMAN LIFE on this planet can be summarized in one word: brutal. Think of driving through life's course of events in either a muscle car or a hybrid. One of these cars may get you to your destination much faster in the short run. But for all that horsepower, the muscle car lacks the stamina, from a fuel-efficiency and an upkeep perspective, to keep moving through the most arduous challenges of life. That stamina comes through for women, who, unlike men, can rely on the genetic fuel of their "silenced" X chromosome.

As I mentioned in the first chapter, just a few short years ago the consensus was that women had the use of only one X chromosome in each of their cells. Just like men. The other chromosome was thought to be more or less completely silenced by the gene *XIST* and turned into an inert Barr body. We now know definitively that is not the case. Like genetic Houdinis escaping the chains of X inactivation, women bring to bear the hybrid power of two X chromosomes to help them survive and thrive. Genes on the so-called silent X are not that silent after all.

Far from being mute, a woman's second X chromosome works tirelessly to help her throughout the life cycle. Genes can escape from X inactivation to help their active sister X chromosome whenever needed. As you're about to see, when it comes to surviving life's challenges, genetic stamina is what counts most.

We are now discovering that of the one thousand genes that are on the "silenced" X chromosome, 23 percent are actually still active. That's a lot of genetic horsepower held in reserve within every one of a woman's cells.

This material, which contains hundreds of genes, is used by every one of a woman's cells whenever she may need it. Using our analogy of a hybrid car, there are times when it's more efficient to use the equivalent of an electric-powered engine rather than the gasoline-powered combustion engine. Having genetic options, as women do, is what counts in life when survival is the goal. The difference here between male and female cells is crucial. Every female cell can call on its genetic reserves of hundreds of important genes in times of need. Men only wish they could do that.

I've been studying the effects of the differential expression and use of genetic material for more than twenty-five years—not just in humans but also in a variety of other organisms, from honey bees to potato plants. What I've learned is that the capacity to respond to and overcome a changing environment or a microbial attack depends on your ability to tap into and use deep genetic resources. It can mean the difference between life and death, and between existence and extinction.

TO TRULY UNDERSTAND the reasons that women outlast men, we need to do a brief but deep dive into the ways in which extreme changes in weather and climate have affected human populations throughout our short history. When the circumstances were at their worst, greater stamina was required to survive what our ancestors experienced day in and day out for thousands of years.

Throughout human history, the only constants across cultures and generations have been birth, death, and hunger. Humans are fully reliant on other organisms for sustenance, which means that we are vulnerable and in a perpetual struggle to eat to survive. Even to this day there is still no city or town anywhere on earth where there isn't someone in dire need of a meal. It's estimated that one in ten people alive today go hungry, with most of those living in the developing world.

No matter where and when, humanity is often suffering from some sort of caloric calamity. This, for the most part, has been historically driven by local weather changes and larger long-term climatic fluctuations. Out of sheer necessity, we have evolved as best as we could to mitigate and survive the worst of famines.

Our ancestors were not picky eaters. They ate whatever food they could find, whenever they could find it. For almost the entirety of our history, we've been eating locally, with our diverse choices linked to the seasons. Our ancestors faced survival challenges from their earliest days, and the odds were rarely stacked in their favor. They had to figure out how to pool their resources in small groups to endure and persist.

More than ten thousand years ago, our circumstances started to change in some drastic ways. We began shifting from purely foraging—procuring all our food from wild plants and animals—to farming and animal husbandry. Producing more of our own food through novel farming and breeding techniques required a much greater effort, especially in the beginning, when any misstep could result in starvation and death.

Agriculture also required us to become more permanently settled. As with all new endeavors, agriculture took a little time and a lot of effort to work out, but eventually our ancestors became immensely successful at farming the land around them, which, in turn, allowed them to store excess food.

Cooking with fire, which we mastered long before the development of agriculture, helped us to further unlock calories in cereal grains and tubers, which, in their native form, were previously beyond our digestive capabilities. All those extra calories eventually meant a lot of extra babies—humans are more fertile when they're not starving. The more successful we were at producing food, the more human babies there were on earth. And subsequently, the more sensitive our population grew to fluctuations in food production, the grander the scale of a possible disaster. So, the cycle began—one that we are still wedded to and will be for the foreseeable future.

Our current global food-supply chain is primarily influenced by one dominant variable: weather. Any sudden changes in the local variables, such as too much or too little rainfall, can result in famine on an epic scale. Providing regular and nutritious meals to billions of humans today is no easy task, and this dilemma is one reason that I have spent many years conducting research with plants such as potatoes, trying to figure out how best to optimize and improve the nutritional features of the foods we produce and consume.

Most of the food available at the local supermarket is made up of a very small portion of what our ancestors ate before the advent of a globalized food chain. People didn't just eat one variety of a carrot or an apple—there were literally dozens, if not hundreds, of cultivars of every type of crop that was grown or harvested.

Our modern problem is not strictly about the quantity of production; it's also about the limited nutritional quality per calorie consumed. The relationships between our dietary needs and our genetics are complex. Given what we've had to contend with in our evolutionary past, we can get by in the short term on food of little or no nutritional quality. Surviving and even losing weight by eating only Twinkies is technically possible—although I wouldn't recommend it.

Studying the genetic capabilities of plants and insects in unforgiving environments often necessitates my travel to parts of the world that are far removed from any tourist site or destination. When I return to a rural site just a few short years after setting up a project, I sometimes barely recognize the natural terrain. Either the soil has been split open by the plow and crops have been planted, or new housing has gone up. Looking beyond our horizon in search of more and more land has been the norm ever since our initial successful attempts at taming life thousands of years ago.

Like all meteoric successes, when it comes to agriculture, ours has been a complicated one. Across the globe, parcels of old-growth forests and jungles have been cleared or burned almost overnight, and many of the world's bodies of water have been emptied of life, all with the goal of providing us humans with more food. I'm amazed by how quickly things change in our attempt to keep ourselves fed. We often ignore just how much we depend on the cooperation of the local, seasonal weather to ensure the perfect parameters of water and sunlight. Even the smallest perturbations can quickly spell disaster.

THE INTERPLAY OF CLIMATE and agriculture is particularly evident in the ruins and monuments left behind in disparate parts of the world. One such spot is about a hundred miles north of Lima, Peru, where the UNESCO World Heritage Site of the Sacred City of Caral-Supe sits. The five-thousand-year-old site contains the remnants of a city complex that's thought to be the oldest center of civilization in the Americas. Its six large pyramids were originally built to withstand earthquakes. Having more or less done that, they are now slowly crumbling back into the desert from which they rose thousands of years ago.

That fact makes the site feel all the more desolate. It looks like everyone who once lived there left in a hurry. The consensus is that, as with so many other ancient sites throughout the world, this city was abruptly abandoned as the climate or local weather changed and the food supply decreased. After all, when the food runs out or the rivers run dry, there's not much choice. Leave or die.

The process of starvation begins rather quickly when we cannot meet the energy demands of the day. Animals the world over have come up with strategies to find the food they need to survive. The most obvious is migration. But loco-moting to calorically rich and plentiful pastures works as a strategy only when greener pastures are actually available.

There is no denying the fact that we humans are a vora-cious bunch of animals. It's almost as if our insatiable appetite is trying to fill a bottomless belly carved out by the past hun-gers our ancestors experienced. That's why the most gluttonous survived. We are also a fickle and forgetful species. There's one thing, though, we've never been able to forgive or forget, and that's hunger. Even if we tried to forget, it wasn't long before another famine came along to retell a very old story. Which is why we should all be excused for our rapaciousness—it's literally been hardwired into our DNA.

Since animals need to eat plants and other animals to sur-vive, they are dependent on the cyclical nature of food avail-ability. This, of course, is tied to local weather and climate patterns. The only way to ensure that you have something to eat in times when food is less plentiful is by embracing the second survival strategy. This involves squirreling it away.

Every organism on the planet has some type of energy-storage strategy. The potato plant stores its energy in the form of starch produced through photosynthesis in its leaves in underground tubers, hidden away from hungry animals. The honey bee, *Apis mellifera*, makes and stores honey—so that it can have something to eat during the winter when flowers are no longer available.

To decrease how much energy we use in persistently lean times, we start to limit the amount of energy we expend as soon as starvation kicks in. We do this by slowing down our metabolism, a survival strategy that both women and men employ. This is one reason that permanent weight loss is such a challenge for everyone. Our bodies literally evolved to become calorically thrifty, and the longer we sense a lack of food coming in, the more our energy expenditures slow down to compensate. This is why the first pounds always come off more easily than the last. With every ounce we shed, our bodies demand increased efforts to continue the weight-loss trajectory.

Like other animals, humans can also store excess food, whenever it's available, inside our bodies as fat. It's what our bodies eat when there's no other food around. Women on average have up to 40 percent more body fat than men of the same height and weight. Women also tend to carry their fat in the femorogluteal region (hips and buttocks). Most men, on the other hand, have a higher percentage of muscle mass. Although useful, having more muscle also increases the amount of energy required just to get by. The body of a man requires more calories to function than that of a woman of identical weight. This can be problematic when there's no food left to eat.

By having less muscle mass, as well as a lower resting metabolic rate, females are genetically endowed with the ability to be more energetically thrifty than males. In times of need, it pays off to carry some extra calories inside you. It's also one of the reasons that women, as opposed to men, survive in lean times.

FARMERS THROUGHOUT HISTORY have realized the importance of producing as much food as possible. Mastering how much you can produce in a short growing season can often be a challenge. Sweden, for example, has long summer days but an overall short growing season. This can result in wonderfully productive years, and yet a few lost days of sun can be disastrous. Famine quickly ensues.

In summer 1771, that's exactly what started happening. Not just in Sweden but in other parts of Europe as well. There were widespread crop failures from abnormal weather. Things didn't improve much the following year, or the next, which resulted in a spike in food prices, which exacerbated the situation. Malnutrition led to the spread of infectious diseases, and dysentery thinned the population out even further. But, as in every other unfortunate time of famine and pestilence in human history, women outlasted men.

What's unique about the crisis that transpired in late-eighteenth-century Sweden is that it happened during a time when there was a virtually complete and accurate vital registration system. It included death registration and census data. Before the crisis that hit Sweden, life expectancy was 35.2 years for women and 32.3 years for men. During the crisis, it dropped to 18.8 years for women and 17.2 years for men. And then after the crisis, life expectancy climbed to 39.9 years for women and 37.6 years for men. With death records in hand for these years of famine in Sweden, we now know, with arithmetic certainty, that females had a compelling survival advantage.

Crises strike, and women have the biological ability and physiological stamina to endure. And it's not just adult women who outlive adult men. Researchers reviewing the accurate vital registration data from eighteenth-century Sweden discovered that infant girls were outliving infant boys. This is the same early-in-life female survival advantage that I witnessed firsthand during my time at the Tarn Nam Jai Orphanage and in the NICU.

As I've explored the genetic basis for persistence and stamina already, I now want to expand the discussion to elaborate on the genetic basis for longevity, which endows human females and many plants, like potatoes, with the fortitude that is required to survive life's continuous stream of challenges.

Plants respond in the same way as humans when they're challenged, relying on their formidable genetic abilities when the going gets tough. Sometimes they do this by storing away their own food source in the form of starch—it's the botanical version of the human process of storing fat.

Long before we started eating them, potatoes were simply meant to be eaten by the potato plant itself, since potatoes contain the stored energy created by photosynthesis during the growing season. Once the season is over, the plant at the surface dies back. But the potato does not simply function as a caloric storage device, the way our fat does. An identical new plant can grow from it using all that accumulated energy stored in its underground caloric food bank. And then the process begins again. By saving some of the energy they produce in one season to use in another, potato plants increase their own odds of survival.

The one thing plants cannot do easily is locomote. Since almost all plants are stationary, they don't have the option, as avian species do, for example, to fly to a different site. That means plants have evolved unique abilities to deal with the stressors of life. What allows them to survive is their ability to respond in kind. Plants do this by continuously tweaking their genetic machinery to adapt to day-to-day changes in their environment.

I've seen how stressing potato plants by limiting their availability to water, for example, can increase the amount of carotene antioxidants they produce. Eating foods that respond in this way genetically can have a positive impact on our own health. That's where the antioxidants from many vegetables and leafy greens come from. We're literally consuming and benefiting from a plant's genetic responses to the stresses of life.

IT WAS THE MIDDLE OF JULY, and I was back in Peru in a region of the Altiplano plateau, bouncing around in the passenger seat of a jeep climbing ever so slowly to an altitude of fifteen thousand feet. I was deep in the heart of the Andes to further study how potato plants genetically respond to and survive the challenges of growing at the top of the world.

It was about a two-hour drive from the ancient city of Cuzco, which left me lots of time to talk with Alejandro, my driver and an expert guide to the farming region in this exceptional part of the world. As we snaked our way higher up the side of a mountain on a very narrow road, I was reminded of what can quickly go wrong by the rusted carcasses of cars lining the valley below.

Most supermarkets in the developed world have at best only half a dozen varieties of potatoes. But outside the modern monocultural world of convenience, there are actually an additional five thousand varieties of potato known to science, most of which originated here in the Altiplano. Every potato anyone has ever eaten can trace its ancestral roots back home to this part of Peru.

I was particularly fascinated by the fact that a certain variety of potato, known as Mama Jatha (mother of growth), thrives in the extreme conditions present at high elevations. What could this teach us about the human capacity to endure harsh situations and environments? If potatoes could respond genetically to survive harsh environments, maybe that data could point us toward a future in which we induce a genetic ability within ourselves to do the same.

Taking one hand off the steering wheel and both eyes off the road, Alejandro enthusiastically pointed out my window. "Look, look—over there . . . Do you see that?"

I shielded my eyes from the blinding sun, which at that altitude was hitting us with an extra 30 percent dose of ultraviolet radiation than what we'd get at sea level. I couldn't help thinking about what all that extra solar radiation was doing to my DNA. I forgot to apply sunscreen before heading out the door that morning, and the extra ultraviolet radiation I was exposed to throughout the day was likely making millions of tiny cuts in the DNA within my skin cells and retinas. We were well over an elevation of thirteen thousand feet at that point. I squinted and tried to focus where Alejandro was pointing. All I saw was a rather small plot of nondescript potato plants, with not much else growing anywhere—not many trees or shrubs. According to Alejandro, these were newly planted fields, at both an elevation and a location he'd never seen before.

Potatoes are not new to this land. They've been under cultivation here for at least eight thousand years and have provided the caloric carbohydrate fuel for many vast, ancient empires, including the Inca.

Domesticated potatoes (*Solanum tuberosum L.*) are just as essential to our survival today as they've always been throughout history. Right now, they are considered to be the most important non-cereal food crop worldwide. The potato is a member of the large Solanaceae family of plants and is a close relative of the eggplant, pepper, and tomato.

Unlike other crops, potatoes do not need traditional potato seeds to cultivate. So, wherever potatoes are planted, actual potatoes are used. These are called "potato seeds"—which are a previous year's crop. If you want to get more potatoes, you just plant potatoes.

In this way, a potato plant will always have the same DNA as its parent, and will look and taste more or less the same. But they will never be identical. That's because potato plants respond to the challenges presented by their location. A plant can have smaller and thicker leaves if it's growing at higher elevations, for example.

Another way the potato plants growing at this altitude respond to this distinct survival challenge is by producing more phytochemical compounds that act as sunscreen for the plants. Having more antioxidants available allows these potato plants to protect themselves from that extra dose of ultraviolet radiation.

"It's not *just* another field of potatoes . . . It's really amazing. I've never seen potatoes growing this high before. This is a first," Alejandro said, as if reading my thoughts. He went on to tell me that the microclimate here has been changing, and this specific area has been warming up, allowing farmers to plant potatoes at higher altitudes. Although more land is available for some farmers because the climate is changing, it's not yet clear if this area will also become drier or wetter as the temperature increases, and therefore whether the farming here will be sustainable in the future. Alejandro mentioned that some farmers have had to move their farms to a lower altitude in areas nearby, as higher elevations were becoming too dry to farm.

Alejandro pulled the car off the road and parked in an empty field so that I could speak to the farmers tending their potato plants. At this altitude, I felt winded just entering and exiting the jeep. Alejandro energetically ran up to the farmers and started busily peppering them with questions in Spanish that I'd prepared in advance. He called me over with a loud "*Oye*, Sharon!"

Alejandro was only one hundred feet away, but deeply immersed in the throes of altitude sickness, I was really struggling to make it over to him. A trifecta of flatulence, a throbbing headache, and extreme shortness of breath came over me. Every step was an immense effort.

Noticing that I was suffering, Alejandro tried to keep me moving and was now dramatically waving me over to him. He was holding two freshly dug and colorful tubers in each fist. "Look, we found them. *Puma maki* potatoes!" That worked. With high-altitude-grown potatoes now at the forefront of my mind, I inched my way over to the field where Alejandro was standing.

Puma maki translates into English as "puma paws," and the

potatoes actually resemble the paws of the large and secretive feline. Alejandro pulled out a pocketknife and sliced into a raw tuber. Beneath the dark purple exterior of the skin was a cream-colored interior speckled with striking linear streaks of more purple.

Holding one of the dark purple *puma maki* potatoes in my hand, I wondered how many chromosomes it might have. Wild potato plants and humans are diploid organisms, which means that we both have two copies of each chromosome in the nucleus of our cells.

Domesticated potato plants, on the other hand, can have a higher "ploidy" number (meaning the number of sets of chromosomes in a cell), depending on the variety. Most of the potatoes on European and North American tables are tetraploid, containing four copies of each chromosome, instead of two. Some potato plants are even hexaploid, meaning they can have up to six copies of each chromosome.

We still don't know for certain why some plants—like potato, wheat, and tobacco plants—have duplicated their chromosomes while others have not. Might there be a benefit? Having extra copies of the same chromosome does result in higher levels of genetic diversity, while also protecting a plant from deleterious mutations. Are there any examples from the vertebrate world of a survival advantage that's directly tied to having greater access to genetic information that aids survival?

Of course. The XX female. This is why human XX females have a survival advantage at every point in the life course over males. Regardless of the specificity of the catastrophe, be it a famine that was triggered by a twentieth-century collectivist political ideology, pestilence spreading from poor living conditions, or an environmental upheaval that renders life close to impossible, more women survive.

This female survival advantage is directly tied to a genetic female's ability to reach far back into the genetic history of humanity and use a variety of tools that males lack. Like polyploid plants, which can draw on more genetic knowledge acquired over millions of years of survival, women just do life better. I believe this phenomenon is the reason why human females are genetically superior to males, as females are diploid for the X chromosome while males are not.

As I've mentioned, we also know that about 23 percent of genes on the so-called "silent" X chromosome in females can escape inactivation and be active in every one of their cells. This gives females access to more versions of those very same genes that they can choose from and use. Having access to different versions of the same genes in the same cell is what I refer to as genetic diversity, and being able to use them is what I have termed genetic cellular cooperation, which gives human females a superior genetic survival advantage.

MY VISIT TRIGGERED the preparation of a mind-boggling and incomprehensible buffet of delectable potato-based epicurean delights. A few hours after I arrived on the Altiplano, an incredible rainbow of botanical potato colors was produced and on display, cooked by the locals and ready to eat. What was most interesting was not just the variety of colors but also the spectrum of flavor, texture, and mouthfeel. These potatoes were like nothing I had ever tasted before.

There was *papa marilla*, which had a surprising grittiness to it, almost like someone spiked it with a little sand. There was also *papa negro*, which had a charcoal-black exterior, a yellow interior, a slightly sweet taste, and a floury texture. I quickly lost count of the multitude of freshly harvested and cooked potatoes on our table, but I tasted as many as I could.

Alejandro reminded me that although potatoes are plentiful here, there is a lack of protein in the diets of some rural Peruvians, especially the youngest, because of the expense of acquiring high-quality protein sources in that part of the country.

We passed a good number of farms on our long drive back to Cuzco. We stopped to speak to more farmers along the way. As we continued our descent and the altitude sickness started to lift, Alejandro pointed out where some farmers have now started planting quinoa and corn instead of just potatoes. "These other crops . . . they can't survive up high," Alejandro explained. "Those crops, those plants, are really not strong enough to deal with all of the stresses of the Altiplano. The cold and often very dry conditions are just too much for them. But these *papas*, these amazing potatoes, are really the master survivors up here. I think that even long after all the people will be gone in the world, these *papas* will still be here. Thriving."

Alejandro was right. Potatoes are survivors and they manage where other plants fail because they can draw on their genetic strength and history in times of crisis. Responding with resilience to environmental challenges like less water and too much or too little sun endows them with the ability to outlive other plants that would simply perish. Potato plants can also rely on their underground storage system of tubers for food, which allows them to survive until the environment is better suited for their growth.

As the climate keeps changing on the Altiplano and beyond, humans, like all other organisms, will also be forced to draw on our own strengths and keep adapting if we want to continue to endure as a species. Being able to access and use more genetic information in every one of their cells, as well as being able to store more energy in their bodies, is what gives human genetic females the superior vigor and stamina to outlive males. Respond, adapt, or die—this has been the hard and cold mantra our species has been living by since our earliest days on this planet. And some of us have proved more successful at surviving than others.

———

AS A GENETICIST AND PHYSICIAN, I've spent a long time thinking about the implications of the genetic differences between males and females when it comes to having the stamina to survive pathological disease processes. I've even experienced the repercussions of these differences personally. This was exemplified by my friendship with Simon Ibell.

A company I founded more than a decade ago was expanding and needed more space. Simon's office was just a few doors down from my new one. And so we got to see a lot of each other. When Simon was a young boy, his doctor predicted that he would likely grow to be over six feet in stature. Simon never made it past four feet, eight inches. But his height didn't matter to anyone who knew him—it was his immense presence and infectious charm that made such an impact on the people close to him.

Simon later told me that it was our company's logo of a fingerprint that initially caught his eye. "You're like a genetic detective looking to capture rogue genes," he remarked one night while we were both working late on our various projects. "What are the chances, then, that I'd be your neighbor . . . that of all people I'd be the one right next door."

My biotechnology company, Recognyz Systems Technology, was developing facial recognition software and cameras—the type used today to unlock smartphones and enter homes—to expedite the diagnosis of rare genetic conditions. According to the latest figures from the National Organization for Rare Disorders, a nonprofit working to raise awareness of rare medical conditions, there are more than seven thousand rare diseases that we know of, and the number keeps growing as we learn more. Although individually these conditions are not commonplace, when you add them all up, more than thirty million Americans and an estimated seven hundred million people worldwide are affected by one today. My company was also creating a mobile app to use in health-care settings that could help families and the medical workers involved in their care to decrease the *years* it used to take to arrive at a proper diagnosis.

"So, what are the chances of your genetic detective agency opening up shop right next to me?" Simon said with a knowing smirk. "I guess my secret is going to finally come out, and I'll have nowhere else to hide." Simon's ability to make light of his situation was just one of the many reasons everyone who knew him couldn't help loving him.

Far from hiding, Simon was working hard. He had devoted years of his life to spreading the word and raising money through his nonprofit iBellieve Foundation for research toward finding a cure for Hunter syndrome, also known as mucopolysaccharidosis type II (MPSII).

Affecting only 1 in every 3.5 million people, Hunter syndrome is not by any means common. So, the challenge faced by Simon, like all those who advocate for people with rare diseases, was to figure out how to motivate individuals to help find a treatment or cure for a condition that affects so few. And yet Simon, always inspired, used to say, "You just have to get people to believe for the smallest moment that change is possible . . . That's when the miracles happen."

When it came to miracles, Simon had it covered. As a child he was diagnosed with Hunter syndrome and given a life expectancy of only a few more years at best. Simon not only outlived every revised life-expectancy estimate he was given, but he even outlived some of the doctors doing the prognosticating.

I knew from my work in rare diseases that Hunter syndrome is an X-linked genetic condition. So as we've seen, for the most part, only males, lacking a backup X, are affected. Without a spare tire in the trunk of your car, you're not going anywhere fast if you get a flat.

A housekeeping gene called *IDS* that resides on the X chromosome wasn't functioning properly in Simon, and consequently, the enzyme that was normally made from *IDS*, called iduronate-2-sulfatase, wasn't going to be working as well either. Imagine trying to put together a new piece of IKEA furniture while missing three key pages from the deceptively simple instruction manual.

If you don't have enough iduronate-2-sulfatase in each cell, you're going to develop a waste disposal and recycling problem fast. Why? Because this enzyme helps break down and clear out cellular waste products from within the cell.

If the enzyme malfunctions or you don't have enough of it, before you know it, your cells will be overloaded with waste products and your organs will begin to expand. Children with Hunter syndrome can end up with enlarged hearts, livers, and spleens, which put pressure on their small chests, causing constant excruciating pain. There's simply not enough room for their expanding organs.

It's most likely that Simon inherited his *IDS* gene that wasn't working well from his mother—as it lives on the X chromosome. Why was Simon affected with Hunter syndrome while his mother wasn't? The answer has to do with cellular cooperation: his mother, Marie, is alive and well today because her cells cooperate.

As we know, females have the use of two X chromosomes, which helps when one of them has a mutated gene. Far from having only a mere backup copy with genes escaping X inactivation, a woman's cell can also provide the genetic help that's needed to keep a sick "sister cell" alive that otherwise would have perished. Male cells like Simon's don't have that option. Marie's cells that are using an X that can't make iduronate-2-sulfatase are being kept alive by her other cells using a different X that can. By cooperating, Marie's cells can outlast Simon's, even though they both inherited the same mutation on their X.

When I first met him, Simon was already receiving infusions of idursulfase, which was one of the most expensive drugs on the planet at the time—a year's worth cost around three hundred thousand dollars. And on top of that, it's far from a perfect drug. It's effectively the same enzyme that Simon's *IDS* gene couldn't produce on his own since it lacked the right instructions. The problem with this drug is that idursulfase cannot get into every cell in every part of the body.

Marie didn't have Hunter syndrome and didn't need to take idursulfase because her cells could make enough of the enzyme to share with the cells that lacked it. This is often referred to as genetic redundancy, but that's not really correct. By sharing enzymes between cells, women can keep some of their cells alive that otherwise would have died. One cell secretes the enzyme and the deficient cell takes it up through a process known as mannose-6-phosphate-mediated endocytosis. This type of cellular cooperation salvages cells that otherwise would have died but that still have other useful versions of different genes on their own X chromosome.

Indeed, cellular cooperation between cells using different X chromosomes is one of the major reasons for the superiority of genetic females. Imagine two cells side by side, each using a different X chromosome and being able to share gene products between them. Marie's other X chromosome made the enzyme from a working copy of the gene and shared it with the cells that didn't have it.

So, right off the bat, women are genetically superior—they have a backup copy of the X, *and* they can cooperate and share their genetic wisdom between cells to combat genetic deficiencies, which can literally mean the difference between life and death.

For people like Simon, starting the drug as early as possible in life is crucial because it can delay some of the symptoms of the milder forms of Hunter syndrome. Unfortunately, it can't prevent or reverse the characteristic heart enlargement and, for some people, the neurological deterioration.

The immense depth of Simon's character crystallized for me when I saw him after he had a meeting one day with a family who had just received a diagnosis of Hunter syndrome for their tiny eighteen-month-old boy. Simon knocked on my office door and came in, breathless and excited. Part of Hunter syndrome results in breathing difficulty due to obstruction of the airway, and so Simon had some days that were better than others. As he caught his breath, he told me that this young boy could now start the drug idursulfase early, and that maybe it would make a world of difference, letting him live long enough so that he could receive a new life-changing therapy down the road that would work even better. In all his recounting of this boy's story, I never once picked up on a trace of regret from Simon that he himself wasn't able to take idursulfase as a child.

What Simon's life illustrates is that the payout for having the use of more than one X chromosome is often survival. No matter how hard they try to overcome the challenges of life, males always begin theirs at a genetic disadvantage.

The last time I saw Simon, he was his usual optimistic self. He filled me in on everything he was up to—including a new romantic relationship that he was very excited about. Simon passed away in his sleep on May 26, 2017—he was only thirty-nine years old. His mother, Marie, is still alive today.

—

WE SEE A COMPARABLE genetic superiority phenomenon in birds—worth mentioning because they use a similar chromosomal sex-determination system as humans, but in reverse. In this way, male birds are like female mammals, in that they have the parallel use of two X chromosomes, which in birds are called Z chromosomes. The equivalent to the

Y chromosome, which female birds have, is called the W chromosome.

With birds (the modern-day descendants of dinosaurs), males tend to be the stronger genetic sex and don't have the equivalent of human male X-linked diseases. Female birds, on the other hand, are like human males in that they bear the brunt of X-linked conditions, which in birds are Z-linked. Male birds are the ones that tend to live longer. Indeed, the same holds true for lizards and amphibians—the stronger genetic sex is the one that inherits the equivalent of two mammalian X chromosomes.

I came to learn about male bird longevity quite by accident after meeting Chef Yoshihiro Murata. While I was conducting research in Japan, I was fortunate to experience a *kaiseki* meal with Chef Murata at his restaurant in Tokyo. *Kaiseki* is a special type of Japanese cuisine that is highly seasonal and exemplifies *washoku* (和食), which has been designated by UNESCO as an intangible cultural heritage that expresses Japanese people's respect for nature.

The food was exquisite. With his four restaurants and seven Michelin stars, it's easy to understand why Chef Murata's gustatory marvels are held in such high esteem. The meal consisted of a seemingly endless stream of courses—actually fourteen in all. The next day over tea, Chef Murata and I spoke about his latest project, a multivolume set of books about Japanese cuisine. When I asked him how many volumes he was planning on including in the series, he smiled and answered with his signature dry humor, "Many."

He showed me what was to be the first introductory volume, along with some photos he was considering including in future books. I asked him what other special dishes I should try eating while I was in Japan. He told me to try sweetfish (*ayu* in Japanese) and showed me a photo. I had eaten them before, but there was something about these particular *ayu* that caught my eye. Both fish in the photo had what looked like two distinct perpendicular marks about halfway down their bodies.

Chef Murata noticed me puzzling over the unique marks and explained, "It's from a bird beak . . . the one who catches the fish." He proceeded to pantomime the entire process.

The fish weren't caught in a net or raised on a farm. Rather, this particular *ayu* was the result of a centuries-old custom that has now practically disappeared. At first blush, I just thought this story was an elaborate joke at my expense, but from the somber look on Chef Murata's face, I understood that he was serious. I had to find out more.

The following week I went to meet the birds who catch *ayu*. While I was sitting in fisherman Shinzo Yamazaki's boat, a beautiful cormorant, with striking emerald eyes, a black body, white cheeks, and a splotch of mustard color beneath its beak, stared at me, looking suspicious. Yamazaki told me, "Not to worry—this bird doesn't eat people." As we pushed off, Yamazaki placed a rope around his bird's chest, getting him ready. He pointed to the metal ring around the bird's neck and explained that this was what stopped his bird from swallowing the *ayu*. He nudged the bird into the water, and a few minutes later it was back, its neck bulging.

Yamazaki gently opened the bird's mouth and squeezed the protrusion, and out came three small fish. I felt a little sorry for the shortchanged cormorant. As if reading my mind, Yamazaki reached into a box nearby and gave his bird some ground-up eel as a reward.

The practice of using cormorants to catch fish is thought to have been introduced to Japan from China in the seventh century, perhaps earlier. Yamazaki has six birds and would like more, but his wife complains that they take up too much room in the house as it is. He also tells me that he prefers male cormorants, even though they are more expensive, as he finds them healthier and longer-lived. As with human females, male birds on average have been found to live longer. This shouldn't be a surprise by now. After all, male cormorants are endowed with the equivalent use of two X chromosomes, just like human females. It's not only life insurance companies that know and take advantage of the difference in life expectancy between the sexes. With all the work and expense that goes into training these birds to fish, it's no wonder that having one that lives longer pays off in the end.

———

IF YOU TRIED TO CONCENTRATE all life's grueling physical challenges into just one group of sporting events, it would look like the obscure domain of ultra-endurance competitions. In this world, Courtney Dauwalter is something of a rebel.

She won the Moab 240, a 238.3-mile footrace, in 2 days, 9 hours, and 59 minutes. The race is a massive loop course that works its way through Canyonlands National Park, in Utah. Dauwalter was markedly faster than any of the men she raced against, winning over Sean Nakamura, who placed second, by more than ten hours. These are accomplishments that no one would have thought possible even a few years ago.

Dauwalter is not alone in breaking new ground and records as more women start competing in ultra-distance events. Especially in the races that favor sustained endurance and stamina versus short bursts of sheer muscle power alone, something interesting is happening. Women are competing and winning.

Dauwalter lives by her own rules. Fueled by M&Ms, Lucky Charms, jelly beans, and burgers, she certainly doesn't follow the nutritional trends you'd expect from such an elite, high-performance athlete. Eschewing the more traditional types of training, she continues following her own course when it comes to how long and far each of her training runs should be. She doesn't always plan out her runs either: "Sometimes when I leave my house I don't even know if I'm going to go out for 45 minutes or for 4 hours. Basically, I'm just listening to my body and relying on the fact that I feel like I can read the signs my body is giving me pretty well and just going with it."

One thing is for sure: Dauwalter loves to run. Some might even call her extreme—in a hundred-mile race called the Run, Rabbit, Run, she found herself temporarily blinded during the last twelve miles. She still managed to finish.

The Montane Spine Race is another brutal ultra-endurance footrace. It combines a 268-mile nonstop marathon with hilly terrain that involves climbing a total of 43,000 feet (in comparison, Mount Everest stands at a height of 29,000 feet). If that weren't bad enough, the Montane Spine Race is undertaken in the middle of winter, with two-thirds of the course run in complete darkness. All competitors have to carry their own kits and supplies, with no personal support teams aiding them on the course. Sleeping for only 3 hours during the entire competition, Jasmin Paris won the Montane Spine Race in 83 hours, 12 minutes, and 23 seconds. Paris not only was the first woman to win the Montane Spine Race, but also beat the previous race record set by Eoin Keith by an astonishing 12 hours. Paris even managed to find time to express breast milk for her 14-month-old daughter at 4 out of the 5 checkpoints on her way to the finish line.

Overall, men have larger hearts, more lean muscle mass, and an increased ability to get oxygen to where it's needed in the body. But these advantages can be costly. Just ask Rebecca Rusch, the seven-time world-champion mountain biker who has competed against men for more than twenty-five years. As she says, "All these guys will go out hot, and hours later I catch them. They always ask, 'Why do you start so slowly?' And I answer, 'Why do you finish so slowly?'"

It seems like the more arduous the sport, the more

women's genetic advantage of stamina propels them ahead of their male competitors. Exemplifying this trend, German athlete and medical student Fiona Kolbinger recently beat more than 200 men from a field of 256 riders to win the Transcontinental Race. This punishing 2,500-mile ultra-cycling event spans the width of Europe, requiring riders to traverse a paved pass through the French Alps at 8,678 feet, all while being exposed to unpredictable weather. Kolbinger beat out the second-place finisher, Ben Davies, by a comfortable 7 hours, crossing the finish line in only 10 days, 2 hours, and 48 minutes. After winning, Kolbinger remarked, "When I was coming into the race, I thought that maybe I could go for the women's podium, but I never thought I could win the whole race."

Traditionally, it was assumed that men were the stronger sex, but when you look at the numbers, why are baby girls in the NICU obviously stronger than baby boys, and why do more women survive harrowing famines than men? Even when environmental and behavioral differences are taken into account, mortality is always higher in males.

Robust genetic diversity from the use of two X chromosomes and cells that cooperate make all the difference for females. This extra chromosomal diversity and stamina is what endows every genetic female with a survival advantage.

Having two X chromosomes allows women to endure, overcome, and thrive better than men on average, regardless of where in the world they are born and into what circumstances. If there's anything we can learn from the past, it's that when challenges crop up, the levy of flesh will never be equal between the sexes.

When it comes to the ultramarathon of life, there is definitely one sex that continues to dominate.

5

SUPERIMMUNITY: THE COSTS AND BENEFITS OF GENETIC SUPERIORITY

S MALLPOX HAS EASILY been the source of some of the greatest human suffering over the last few centuries, killing hundreds of millions of people in a very short time. Native Americans were especially ravaged, seemingly overnight, through the machinations of an invisible viral enemy.

The Intensified Smallpox Eradication Campaign was launched in 1967, under the auspices of the World Health Organization (WHO), in an effort to snuff out the smoldering embers of this infectious disease. At the time, almost three million people were still dying every year worldwide, and millions of infected survivors were left scarred and disabled for life as a result of the viral scourge. The WHO was determined to change that. And it did. The campaign against smallpox resulted in the first-ever successful, and intentional, worldwide eradication of a human pathogen.

The dime-size scar on my upper left arm is a telltale sign of the smallpox vaccination that I received as a young child. Overall, immunizations against infectious diseases such as smallpox have prevented more deaths and alleviated more suffering than any other medical treatment developed to date. What my vaccination scar represents is one of our greatest collective achievements at making this world a much more hospitable place.

Prevention of smallpox today may be difficult to appreciate when it's no longer savagely killing and maiming millions. Few people are left who can recount what being infected actually felt like.

Here's how it's been described. It begins rather innocuously, with a two-week incubation period during which people usually don't feel that anything significant is wrong. Then the flu-like symptoms begin, often with a high fever and body aches, sometimes accompanied by vomiting that lasts between two and four days. Then a rash starts to form, usually on the tongue and in the mucous lining of the mouth, nose, and throat. After moving to the surface of the face, the rash then marches onward to the arms and legs, before finally making its way to the tender skin of the hands and feet. About four days in, the rash gives way to sores filled with a turbid and opaque fluid. Each sore is tight and painful, and they cover the infected victim from head to toe. A smell of rotten flesh now emanates from the infected. Around the sixth day the sores become hard pustules and start to feel like pearls hidden under the skin. This stage lasts about ten days. The pustules eventually start to crust over, and hundreds of scabs then cover the body.

Not everyone was lucky enough to survive. Death was painful and far from imminent for those with pustules that didn't heal. Their organs and internal tissues would start to hemorrhage and liquefy, leaving victims looking like they'd been mummified, all while still alive. This dreadful process could last for four weeks.

For those who cheated death, when their scabs eventually healed, the skin was left devastated and disfigured with dreaded scars—and many of the survivors lost their vision as well. Pockmarked flesh acted as a visual reminder, causing victims to be shunned by those around them. The one thing survivors did have going for them was that they were almost assured a lifelong resistance to reinfection—although no one at the time quite understood why.

It was only in 1973 that the first grainy picture of a fragment of an antibody—a specialized protein made by the body to fight infections—was published. It was antibodies that were instrumental to our finally winning the war against smallpox.

In 1980 the WHO officially declared smallpox eradicated. Humanity could finally exhale, as a perpetual fear was put to rest. Like billions of other people, I was spared being disfigured and dying from a smallpox infection from the inoculation I received as a baby. I was one of the last people on earth to be vaccinated during the final eradication efforts.

So how did we get here, and what does smallpox have to do with women and their genetic superiority? We conquered smallpox by triggering and then harnessing the latent power of the immune system, one of the most sophisticated biological systems in our bodies. And as we'll come to see in this chapter, genetic males can rarely compete with the immunological arsenal wielded by genetic females.

THE GRAND NARRATIVE of the outstanding scientific achievement of eradicating smallpox often begins with an introduction to the eighteenth-century British physician Edward Jenner. Every microbiology or medical school student in almost every country in the world is taught more or less the same story about the father of immunology: Dr. Jenner is given the starring role of the medical hero who bursts on the scene by discovering how to prevent smallpox through vaccination.

What Jenner was actually famous for *before* his work on vaccination was studying the nesting habits of the common cuckoo bird. The cuckoo lays its egg in another bird's nest, offloading its parental responsibilities on an unsuspecting new parent. During Jenner's time, people thought that the adult cuckoo bird was taking this parental divestment one step further (and in a rather gruesome way) by getting rid of the other eggs and chicks in the nest to ensure that its child would enjoy full access to the food and resources provided by the oblivious surrogate. Through careful observation, however, Jenner determined that the cuckoo parents were not at fault—it was the cuckoo chick that had the homicidal tendencies. The baby cuckoo quickly dispatched all the other eggs and chicks by throwing them out of the nest. For his work on cuckoo birds, Jenner was elected a Fellow of the Royal Society, one of the highest honors a scientist of his time could receive.*

There are a few stories about how Jenner came up with the idea of vaccination for smallpox. One version has Jenner's aha moment occurring while he was still training as a country doctor in Berkeley, Gloucestershire. While there, he overheard a milkmaid say that she wouldn't catch smallpox because she had already had cowpox.

Both cowpox and smallpox are caused by related and yet distinct viruses, with the former evolved to infect cows, and the latter specializing in infecting humans.† Cowpox was an occupational hazard in those days for people sharing an awful lot of time and space with cows.

A different version of the same story states that Dr. Jenner had a patient by the name of Sarah Nelmes, a dairymaid who

---

*It took 150 years after Jenner's death for the photographic proof to materialize that his theory about the murderous cuckoo chick was correct.
†Cowpox and smallpox are caused by two different but related orthopoxviruses.

developed a strange rash on her hand. Having seen many of these infections in women who were milking cows, he made the astute diagnosis of cowpox. Hearing from Sarah that she was immune to smallpox after her infection with the much milder cowpox, Jenner thought to put this idea to the test.

So, Jenner used James Phipps, the eight-year-old son of his gardener, to see whether being infected with cowpox protects against smallpox.* Extracting the pus from Sarah's cowpox-infected hand, Jenner then transferred it to the boy by breaking the natural protective barrier of his skin. Just a few days later, James fell ill with cowpox. Jenner hadn't yet proved that recovering from a cowpox infection provided any immunity to smallpox—to further test his idea, he would have to wait patiently to see whether the cowpox infection would also protect James from smallpox once naturally exposed.

Or Jenner could directly inject James with smallpox to move things along. Jenner chose the latter. Mercifully, James survived this intentional exposure to smallpox. Jenner named his technique *vaccination* after the Latin word *vaccinus*, for "from a cow."

Even though he was ridiculed for it, Jenner continued with this work, repeating his successful vaccination experiment with other children. But it wasn't just that Jenner himself was ridiculed. The father of immunology had his work rejected by the *Philosophical Transactions of the Royal Society* in 1796,

---

*There continues to be widespread debate about the ethics of the use of human subjects in medical experimentation in the historical development of smallpox variolation and vaccination procedures. While no consensus has been reached, the conversation is still valuable and relevant today.

the leading peer-reviewed journal of his time. This wasn't in error. The president of the Royal Society, Sir Joseph Banks, personally did the honor on the advice of two reviewers who read and commented negatively on Jenner's work.

He finally published his work, titled *An Inquiry into the Causes and Effects of the* Variolae Vaccinae, *a disease discovered in some of the western counties of England, particularly Gloucestershire and Known by the Name of Cow Pox.* It was self-published at his own expense in 1798.

Eventually, physicians and their patients started to come around to the benefits of using Jenner's approach to prevent the dreaded pox. Ultimately, he was even awarded grants from the British government totaling thirty thousand pounds (more than one million U.S. dollars today) to continue his important scientific work. But Jenner didn't enrich his own finances with his discovery. Quite the opposite—near his home Jenner built a hut, called the Temple of Vaccinia, where he vaccinated those who couldn't afford to pay for the procedure.

As Jenner correctly presaged shortly after publishing his research, "The annihilation of the Small Pox, the most dreadful scourge of the human species, must be the final result of this practice."

IT TOOK JUST TEN SHORT YEARS after Jenner's initial experiments for tens of thousands of people to be vaccinated. But like many genesis stories of scientific breakthroughs throughout history, this story has an alternate beginning. What I wasn't taught during college (or any of my medical training for that matter) was the important role played by Lady Mary Wortley Montagu in the development of vaccination.

Lady Mary Montagu was born on May 26, 1689, into an aristocratic family. She grew up in London in the manner typical for a woman of her position. Except Lady Montagu was anything but typical. From the time she was a child, it was obvious to those around her that Lady Montagu had an extraordinarily independent and curious mind.

It may not be all that surprising that someone of such indomitable spirit would flatly reject the marriage that her father, the Marquess of Dorchester, had arranged on her behalf. Carving out her own path in marriage against her father's wishes, Lady Mary eloped in 1712 with Sir Edward Wortley Montagu. In so doing, she changed the trajectory of not only her own life but perhaps our world as well.

A few short years after getting married, Lady Montagu was infected with smallpox, from which she recovered. Every time she held a mirror up to her face, her pockmarked flesh was a painful reminder of the devastation caused by this disease. Even her eyelashes fell out during her infection, and they never grew back. Just eighteen months after she recovered, her brother Will came down with smallpox. He wasn't as fortunate as his sister: he died shortly after becoming infected, at the age of twenty.

In the beginning of 1717, Lady Montagu left England behind and accompanied her husband to Constantinople, where he had been newly appointed as the ambassador to the Ottoman court. She immersed herself in the local culture, learning to speak both Greek and Turkish. Of the many things she witnessed and observed at the time, there was one particular local custom called engrafting or variolation that grabbed her attention.*

In a letter, Lady Montagu wrote:

A propos of distempers, I am going to tell you a thing, that will make you wish yourself here. The small-pox, so fatal, and so general amongst us, is here entirely harmless, by the invention of engrafting, which is the term they give it. There is a set of old women, who make it their business to perform the operation, every autumn, in the month of September, when the great heat is abated. People send to one another to know if any of their family has a mind to have the small-pox; they make parties for this purpose, and when they are met (commonly fifteen or sixteen together) the old woman comes with a nut-shell full of the matter of the best sort of small-pox, and asks what vein you please to have opened. She immediately rips open that you offer to her, with a large needle (which gives you no more pain than a common scratch) and puts into the vein as much matter as can lie upon the head of her needle, and

---

*The term *variolation* comes from the Latin word *varus*, for "pimple." Other terms such as *engrafting* and *inoculation* are synonymous.

after that, binds up the little wound with a hollow bit of shell, and in this manner opens four or five veins.

What Lady Montagu probably didn't know was that engrafting was not native to Constantinople. Chinese physicians were already variolating their patients with powdered scabs two hundred years prior to Lady Montagu's reporting.

Lady Montagu's vivid description might sound familiar. That's because the technique is the same process of vaccination that Jenner employed many years later, with one very important distinction: Jenner's vaccination technique involved infecting people with material containing cowpox, not smallpox. Using cowpox was much safer than using smallpox.

Vaccination and variolation both employ the human body's adaptive system of protection, stimulating the necessary immunological defenses to fight off infectious agents, like smallpox. It is precisely this system that women employ more effectively over the course of their lives than men do. The idea behind vaccination and variolation is to elicit a milder infection, one the body can overcome, which then results in some degree of protective immunity. Females, when immunologically provoked, have the ability to respond more forcefully than men to fight challenges induced through immunization.

Apparently, Lady Montagu was so taken by the potential of variolation to prevent smallpox that she had the procedure performed on her son, Edward, in front of the surgeon of the British embassy Charles Maitland. It worked, and her son never developed a full-blown case of smallpox.

In a letter written soon after her son's inoculation, she declared, "I am patriot enough to take the pains to bring this useful invention into fashion in England, and I should not fail to write to some of our doctors very particularly about it, if I knew any one of them that I thought had virtue enough to destroy such a considerable branch of their revenue, for the good of mankind."

After returning to England in April 1721, Lady Montagu made good on her desire to help introduce variolation to her fellow citizens by trying to raise interest in the technique. Being both a woman and someone attempting to usher in a new unknown medical procedure from the East to a conservative medical establishment unfamiliar with it was not going to be easy. It's not surprising, then, that variolation was not embraced by London's medical establishment at the speed that Lady Montagu had hoped.

When another epidemic of smallpox swept through London in 1721, Lady Montagu wanted to have her four-year-old daughter, also named Mary, variolated. She asked Charles Maitland to perform the procedure, as he had witnessed the inoculation of her son back in Constantinople a few short years prior. He refused.

For physicians at the time, and many others as well, it's understandable that cutting someone's healthy vein open and placing therein the pus extracted from a person sick with smallpox would seem bizarre. Also, it wasn't clear to Maitland or anyone else the best technique to employ when variolating. Should one incise a large vein or a smaller one? How much pus from a smallpox-infected person should be used during the procedure? Maitland wasn't being fickle in resisting Lady Montagu's request to variolate her daughter. Given the dangers inherent in variolation—as 2 or 3 percent of patients did develop fulminant smallpox and die—he didn't want to be responsible for inadvertently causing her death.

In light of all these uncertainties, it's understandable that Maitland had more than a few reservations about inoculating the little girl. But Lady Montagu believed it was worth a try. She knew only too well what the alternative looked like: dying or at best being permanently disfigured from the pox. So she persuaded Maitland, who went ahead with the procedure in front of two witnesses. Lady Montagu's daughter fared well after the inoculation, and interest in variolation began to grow—this time from the royal family.

Maitland was given a royal license to perform an experimental trial of variolation on August 9, 1721. In eighteenth-century Britain, capital punishment was still in place. Having access to those condemned into the hands of an executioner for committing seemingly petty crimes created a unique situation for Maitland. It gave him his first test subjects.

In exchange for the possibility of avoiding execution—that is, if they were lucky enough to survive Maitland's experiment—six convicts were treated with variolation. Survive they did.

Variolation worked, just as Lady Montagu had predicted. One of the inoculated convicts was even exposed to a symptomatic and still-infectious smallpox patient to see if he would be immune. And he was. This convict didn't become ill, escaped the hangman's noose, and was pardoned along with his fellow condemned variolated brethren.

In pure Dickensian fashion, Maitland next performed variolation on orphaned children from the parish of St. James, who thankfully also survived. With the evidence in hand regarding the safety of variolation and its protection against smallpox, Maitland was granted permission to now perform the procedure on Amelia and Caroline, the daughters of the Princess of Wales. Both children survived.

A few articles were published in the *Philosophical Transactions of the Royal Society* around the same time that recounted other practitioners' clinical experiences with variolation. The Royal Society of London also received two letters, one in 1714 from Emanuel Timoni and another in 1716 from Giacomo Pilarino, that both described the very same variolation in Istanbul that Lady Montagu had witnessed. Yet it was the attention that was raised after the royal children's variolation that served as the turning point for its acceptance, one that Lady Montagu had fought so desperately for.

Problems with variolation persisted, such as the risk of acquiring a full-blown smallpox infection that would result in disfigurement, or worse. With time, some of the initial uncertainty was overcome with a newer technique called the Suttonian method—which involved making a smaller incision site and administering a smaller quantity of pus. Using the technique first developed by his father, Daniel Sutton (who was neither a physician nor a surgeon) inoculated twenty-two thousand people between the years 1763 and 1766, with only three deaths. This refined method resulted in an obvious and very significant lower morbidity and mortality in those being variolated.*

Returning once again to Jenner and the vaccination scar on my arm, it's easy to see why using a different but related virus to smallpox had many advantages. Cowpox, which evolved to infect cows and not people, was nowhere near as dangerous as smallpox. So, using cowpox over smallpox for variolation was a grand leap forward. Even a less skilled practitioner could still perform vaccination and not worry about killing the patient.

Yet Jenner's work would not have been possible without the tireless petitioning of Lady Montagu. Here's why: Jenner himself had been variolated with smallpox as a young child—otherwise, he may not have lived to be remembered as the father of anything. His fate could have been that of King Louis XV of France, who died of smallpox in 1774. The French were avid anti-variolators. As a witness to the French court observed, "The air of the palace was infected; more than fifty persons took the smallpox, in consequence of having merely loitered in the galleries of Versailles, and ten died

---

*Variolation with smallpox itself can have the opposite effect of what's intended. Even with the Suttonian method, a fraction of those variolated with smallpox would still develop a full-blown infection and die.

of it." After the death of King Louis XV, his grandson Louis XVI ascended to the throne, along with his wife, Marie Antoinette. It was only after the French Revolution, at the end of the eighteenth century, that the French finally began to take up vaccination.

The English were fervent anti-vaccinators at the time, resisting Jenner's new procedure. It certainly wasn't concerned parents who were resisting the much safer protection that vaccination offered. It was the English variolators themselves who were reluctant to give up what had become for them a reliable cash flow.

Both vaccination and variolation require the presence of a specialized part of the immune system, which is called the B cell. As I've discussed, genetic females have B cells that can make not only more antibodies but better-fitting ones as well.

We make new kinds of antibodies every single moment of our lives. B cells that make antibodies use identical receptors on their surface that are shaped just like the antibody they make. They do their work mainly by reacting to a unique and specific shape of the immunogen, which triggers them to become activated. B cells carry about one hundred thousand identical copies of this antibody on their surface, like antennae, as they await that perfectly fitted immunogen, which activates them to start producing their antibody.

So, if a B cell encounters an immunogen and sufficiently binds to it—bingo! Now this B cell will start to divide into daughter cells. Taking about eighteen to twenty-four hours after its receptor has been triggered, this B cell, along with all its daughter cells, will begin to produce and pump into circulation millions of identical antibodies.

There's also meritocracy at work. B cells that have made an antibody that was successful in clearing a pathogen are promoted upward through the ranks and retained in case of a rematch with the same microbial infection. That's why some of the daughter cells are chosen to become memory cells that will be maintained for years in anticipation of a counterattack.

We employ this system every time we immunize some-one. It's why an immunization can provide protection for years—and sometimes for life. We are keen collectors of immunological memories of all our past infections. By vacci-nating, humans encourage and allow immunological memo-ries to form without getting very sick. For the most part, it's worth the pain.

Some of these memory cells or their progeny may be as old as you are. This phenomenon is linked to the reasons why young children are sick so often. Their immune system, like the rest of their bodies, is still developing. With time, enough immunological experience is acquired through life's little and sometimes big microbial bumps that it creates an immuno-logical repertoire vast enough to deal with the potentially millions of invading microbes. This immunological memory allows us to eventually respond faster and more aggressively to any threats—especially the second time around, on a viral or bacterial rematch.

Immunological memory can be the difference between life

and death. Like neurons in the brain that encode past events and skills that we rely on to survive, our immune system uses antibodies that match invaders to remember and kill them upon their return. This is called the adaptive response of the immune system. And when it comes to immunological memory, compared to males, a genetic female doesn't easily forget. Women usually experience more pain and side effects from vaccination injection than men do, but that's actually due to their superprimed immune systems aggressively reacting to the vaccine more efficaciously.

Even though most of us are born with the ability to make our own antibodies, genetic females are much better at it. As I've outlined, women are more capable than men of making better-fitting antibodies through the process of somatic hypermutation, whereby the B cells undergo cycles of genetic mutations to improve their performance. What further distinguishes females immunologically is that their memory B cells (which make specific antibodies) stay in the body for many more years compared with men's. That's why women tend to respond so much better to vaccinations. Women's immune cells literally never forget.

What Lady Montagu could not have known was that there might be a difference in how the immune systems of males and females respond to and remember variolation or vaccination. From an immunological perspective, women also hit microbes harder and faster the next time they appear.

It is immunological memory that caused Lady Montagu to advocate for variolation, even though she may not have understood all the biological details at the time. Without the body's innate ability to make very specific antibodies, neither vaccination nor variolation would be effective. When it comes to making and retaining antibodies, genetic females dominate.

▭

GEOGRAPHICALLY SPEAKING, the cities of Atlanta, Georgia, and Koltsovo, in the Novosibirsk region of Russia, are worlds away from each other. But they have something in common: the two cities hold the last two batches of the smallpox virus.

Since the World Health Organization declared the world free of smallpox in 1980, there has been a lot of talk but little action when it comes to destroying the last batches of the disease. We've managed to do the seemingly impossible—rid the entire world of one of history's deadliest viral diseases. But after everything we've been through with smallpox over the years, breaking up entirely has been hard to do. Like a shoebox with keepsakes from a teenage romance that we can't bring ourselves to part with, samples of this virus still exist, locked far from view.

There is a good reason for this reluctance to destroy them. You never know when they may come in handy. Especially if you need to create a *new* vaccine against smallpox.

It's not only viruses that are kept on ice, under lock and key. Some of my own research has involved developing an antibiotic treatment for the possibility of a weaponized microbe—specifically *Yersinia pestis*, the dreaded cause of the black death and bubonic plagues. Much deadlier than even smallpox, this microbe in the pneumonic form of the plague can kill 90 percent of those infected. The killing potential of *Y. pestis* is amplified, just as it is for many other bacterial pathogens, with a plentiful source of biologically available iron from its host. Young and middle-age genetic males often have the greatest amount of iron stores within their bodies, as they do not normally lose any through menstruation or pregnancy. This can put them at a disadvantage to genetic females when they have to fight off iron-dependent microbes like *Y. pestis*.

For all its killing potential, *Y. pestis* is surprisingly sensitive to existing antibiotics—but that's before it has been weaponized. To make an already deadly microbe into an even more lethal one is surprisingly easy today. If given a small DNA upgrade, *Y. pestis* can become completely immune to all our currently available antibiotics. Giving *Y. pestis* the genetic ability to acquire more iron and at a faster rate from its host—through DNA-editing techniques—can put genetic males, with their iron-rich stores, at even greater risk.

We have to prepare for the fact that a weaponized black death can make its return at any time and begin killing millions unabated if it's released. That's why it's crucial that we always keep some *Y. pestis* microbes in reserve, so that we can tinker with its DNA and develop and test new drugs to fight it. Drugs that hopefully we will never need to use.

The same possibility exists for smallpox. It's possible to

synthesize the virus in the lab. It was after the defection of Soviet scientists during the 1980s that we began to learn how smallpox was being weaponized into even more potent bioweapons. It took until 1992 for Russian president Boris Yeltsin to publicly admit that the Soviet Union did, in fact, have an extensive offensive biowarfare program that included the production of anthrax, smallpox, and plague.

Even without a weaponized virus, there are additional concerns that recent human remains with smallpox or some long-lost tissue sample may be dug up inadvertently, releasing the virus into the world once again. The best way to fight such a virus is still through vaccination, which is why those samples aren't going anywhere anytime soon.

The scar on my left arm was caused by my body's reaction to the live vaccinia vaccine of the New York City Board of Health strain. Unlike both Jenner's vaccine, which used a live cowpox virus, and the one I received, most vaccines today do not use a live virus. To reduce the chances of any adverse effects, we like to use an inactivated or cut-up piece of a virus or bacterium that can't survive and replicate. It's safer this way. But triggering the immune system to respond to a vaccination that isn't as biologically active presents its own challenges.

To overcome this, modern vaccines are often given together with an immunological irritant, which causes an alarm signal to go off in your body and initiates the recruitment of immune cells. This irritant is partly responsible for the soreness we can experience after an injection, which can be much more significant in genetic females because of their superior immunological response.

———

WHAT'S ASTONISHING IS THAT we can make any antibody that we need—even one that has never existed before in any human. Which is why vaccination can work so well. But what happens when you're born without the ability to make antibodies?

Almost all the people who don't have the ability to make antibodies are men. Genetic males born with X-linked agammaglobulinemia (XLA) have a mutation in a gene called *BTK* that's on the X chromosome, and it means that their bodies are incapable of making proper antibodies. Without another X chromosome to step in when needed, males with XLA—as with X-linked colorblindness—are at a genetic disadvantage.

One of the most common illnesses that patients with XLA get early in life is a recurring ear infection. Most of these boys are totally healthy in the first few months of life. That's because they are still full of antibodies given to them passively through the placenta by their mothers during their time in the womb. But those antibodies don't last that long after birth. When those inherited antibodies run out, the real problems for these boys begin.

The treatment for XLA includes recurring lifelong injections or infusions of gamma globulins (another term for antibodies). Gamma globulins are donated and collected from hundreds of individuals and pooled together. They are then infused or injected into the predominantly male patients.

People with XLA are essentially kept alive by the borrowed immunological memories of those who can make and donate their antibodies. But that's not the only thing that's keeping them alive.

The reason males with XLA survive with donated antibodies is that they still have working within them a part of the immune system called the innate response. This part of the immune system is typically the first reaction to a microbial invasion (or to a group of malignant misbehaving cells). The innate response includes what we call barrier defenses, like the skin and mucous membranes, that interface with the outside environment. The innate response is generalist in nature—it's not specific—which is important, because it allows for the rapid response to an invader without a lot of questions asked.

The innate response is run by a group of cells collectively called leukocytes, also known as white blood cells. The chief workhorse of the innate response is a type of leukocyte called a neutrophil.

Since these cells are generalist cells, they use pattern recognition receptors (PRRs). When triggered, these receptors behave like a piercingly loud fire alarm—alerting the rest of the cells in the body that a microbial invasion might be imminent.

Some of the PRRs—such as the toll-like receptor genes *TLR7* and *TLR8*—are found on the X chromosome. Toll-like receptors sit on the surface of immune cells and are used by them to recognize foreign material from invading microbes. Having two different versions of *TLR7* and *TLR8* gives women an enhanced advantage of recognizing microbial invaders. Men are at a disadvantage, having only one copy of each gene. This means that right off the bat, females will have a chromosomally synergistic immunological advantage in responding to a microbe that's trying to invade and get a foothold inside their bodies.

Within only minutes of a breach or an attack, neutrophils arrive ready on the scene, itching for a good fight. Neutrophils can also call for immunological backup, recruiting other cells to help join the fray. Sometimes there's collateral cell damage as the battle drags on and the cells become liquefied—a.k.a. pus.

Of all the billions of neutrophils that are made within us every day in the bone marrow, some are released into the bloodstream while others travel to the liver and spleen. Compared with other cells in the body, neutrophils are not long-lived. Their life span can range from a couple of hours to a few days—but like Pacific salmon swimming upstream to their final resting spot, at the end of their life, most neutrophils head back to the bone marrow, where they commit a form of cellular seppuku, after which they are recycled.

We have a lot of neutrophil cells—fifty billion of them are made every day in our bone marrow. Every neutrophil uses one X chromosome, which means that women have a lot more genetic diversity in them. All the neutrophils in a male are using the exact same X chromosome.

For genetic females, this neutrophil diversity carries over

to other cell types such as macrophages and natural killer cells, which are working hard to eliminate cells that might be infected with viruses or that have become cancerous.

Having both branches of our immune systems—the innate and adaptive responses—working optimally is crucial for our survival. How do we know this? Sadly, we've seen what happens when they are not functioning in cases like David Vetter, who became famously known as the "Boy in the Bubble." For twelve years, David had to live in a protected environment that was relatively germ-free just to stay alive.

The reason that he was the "boy" and not the "girl" in the bubble is that David suffered from an X-linked condition called severe combined immune deficiency (SCID). About half the cases of SCID are caused by mutations on the X chromosome. That's why three-quarters of people with SCID are males. David's doctors performed a bone-marrow transfusion in an attempt to treat him. The bone marrow is full of immune cells, which can fight infections, and the hope was to offer him a cure. Unfortunately, David passed away from lymphoma that was caused by the Epstein-Barr virus—likely accidentally introduced into his body from the bone-marrow transfusion.

David's genetic condition teaches us that, as in the case of X-linked colorblindness, men don't have the same degree of genetic options as women do. Not only that, but when it comes to males and the genes on the X chromosome, their position is one of deficiency when something goes wrong, which it so often does.

As science catches up with the genetic superiority of a female's antibody response, the vaccines we develop will need to take that fact into account. This distinction may necessitate men receiving an extra booster shot of a vaccination or a higher initial dose than what women will require to be equally protected.

IF YOU'RE A WOMAN trying to overcome an infection or a malignancy, immunological overactivity can have its benefits. But this ability can come at a pretty steep cost—one that women alone usually have to pay.

With her millions of fans and a packed tour schedule, things were going really well for the megastar Selena Gomez. But then, at just twenty-two years of age, the former Disney star suddenly found herself feeling profoundly fatigued. Being in the constant spotlight since childhood and ending a very public relationship with Justin Bieber, she had every right to want to take some much-needed time off. There was even some speculation that Gomez had checked herself into rehab.

No one could have known at the time that Selena Gomez was fighting for her life. Her body had declared war against itself and was slowly and methodically killing her, one cell at a time. She wasn't in rehab—she was actually being treated for an autoimmune disorder called systemic lupus erythematosus, which is more commonly known as lupus.

Gomez is not alone. Around five million people worldwide

have the disorder, and women bear almost the entire burden: females are overrepresented to the tune of 90 percent of the diagnosed cases.

Although it was known to Hippocrates by another name, lupus was actually described more than two thousand years ago. The condition, which causes facial scarring, was thought to have gotten its name in English during the fourteenth century, after the Latin word for "wolf." Some people believe that it was named after the wolf because the distinctive facial rash that can develop resembles the color pattern on a wolf's face. Others believe it was named that because the facial scarring that can result from a type of lupus resembles a healed wolf's bite.

More recently, many people affected by lupus have used the metaphor of the wolf to describe the symptoms of this destructive autoimmune condition. As the writer Flannery O'Connor, who had lupus, described it, "The wolf, I'm afraid, is inside tearing up the place." O'Connor eventually lost her own battle with lupus at the age of thirty-nine.

In Selena Gomez's case, her overly self-critical immune cells were mistakenly targeting and trying to kill her normally functioning cells. Beyond that, her overly critical immune system decided to turn against her kidneys by having B cells make antibodies that specifically targeted them, causing an often-fatal complication called lupus nephritis.

Before she knew it, both her kidneys gave out and shut down. Just weeks away from dialysis, her only hope was a kidney transplant, which miraculously came from her best friend, Francia Raisa.

Currently, there are about one hundred autoimmune diseases. The National Institutes of Health (NIH) estimates that more than twenty million people in the United States collectively suffer from them, even though individually some of the diseases are rare. Overall, they are the third leading cause of morbidity and mortality in most of the developed countries around the world. What unites many of them is that they can be chronic and debilitating.

For the most part, autoimmune conditions predominantly affect females—over 80 percent of those affected are women. As the fifth-leading cause of mortality in women, autoimmune diseases are far from benign. So, if women are the stronger sex from a genetics perspective, why do more women suffer from autoimmunity?

Originally scientists didn't believe that the immune system would even be capable of mounting an attack on itself and harming the body as a result. What would be the sense of that? The body attacking itself? Preposterous.

In 1900, Paul Ehrlich, who would be awarded the Nobel Prize in Physiology or Medicine only a few short years later for his pioneering work in immunology, labeled the impossibility of the immune system turning on the body as "horror autotoxicus." Even as Ehrlich said this, reports began to emerge suggesting that in fact this was exactly what was happening.

By the 1950s and 1960s, there was more of a scientific

consensus that several diseases, such as multiple sclerosis and lupus, were in fact caused by autoimmunity. It also became more apparent at the time that these diseases were more common in women. No one could imagine why there would be this diagnostic numerical imbalance between the sexes. Many doctors and the broadly male scientific establishment assumed that women were just more vocal about the pain and discomfort they were experiencing from the symptoms of autoimmune conditions like Sjögren's syndrome, rheumatoid arthritis, autoimmune thyroiditis, scleroderma, and myasthenia gravis. Clinicians thought that men were silently bearing it and not seeking any medical attention and therefore weren't represented in the numbers. The assumption was that when it came to autoimmune disease, there wasn't an actual numerical difference between the sexes.

We now know definitively that that's not the case. Women suffer disproportionately from autoimmune conditions. In fact, almost all autoimmune conditions are overrepresented in females, no matter where in the world they occur.

Classically, we think of autoimmune diseases as part of the adaptive response, with the damage being caused by cells like B cells that target the body by creating autoantibodies. Instead of going after and targeting an invading microbe, we mistakenly target ourselves. Sometimes the antibodies target a receptor on the surface of the cell and block its action—like sticking a wad of gum in a lock so the key doesn't work.

Other times, these antibodies can cause direct damage to

the cells and subsequently the tissues (which is called Type II hypersensitivity). In Type III hypersensitivity, which is evident in lupus, autoantibodies bind self-antigens and clump together. Similarly, clumped-together proteins (a mix of immunogens and antibodies) get stuck in the narrow channels and vessels of the body. If this weren't bad enough on its own, these clumps of cellular and protein material are trapped in tight spaces, which triggers inflammation, exacerbating the situation by causing pain and swelling. This is likely what Gomez was experiencing as she desperately waited for her lifesaving kidney transplant.

The cost of female immunological aggression is a higher risk for autoimmune disease. An immune system that evolved to protect Gomez from invading microbes had gone rogue and turned against her. I believe that the higher level of autoimmunity that happened in Gomez, and that can potentially happen to all genetic females, is the cost and result of women's genetic superiority.

THE THYMUS IS PRIMARILY a lymphoid organ sitting in the chest, just below the neck. While our B cells make antibodies to battle pathogens, we also have T cells, which can specifically help target and even kill invaders directly.* The thymus

---

*There are many different types of T cells. Some of them target cancer cells and cells that are infected with viruses. According to new research, there are also specialized types, such as $\gamma\delta$ T cells, that may be able to kill bacteria directly.

is where the T cells of the immune system's adaptive response go to finish their training.

T cells are born in the bone marrow. When T cells graduate from the bone marrow, they travel to the thymus for a higher level of immunological education. And it's a brutal schooling. Most T cells don't make it out alive—only about 1 percent of those that enter are thought to survive. That's because so many T cells recognize the "self"—our own body—as foreign, and if given the chance, they initiate an attack.

The thymus gland makes amazing contributions to our lives. The organ serves as the key educational system for T cells—without which our T cells would never be civilized. No thymus, no life. But the thymus doesn't beat like the heart, and it isn't big like the liver, and it shrivels as we age, so we don't think much about it.

Especially for women, the thymus is a mixed blessing and a substantial burden. It is a blessing because T cells within a woman's thymus are turned into highly trained aggressive assassins. It is a burden because, more often than not, those same killing abilities are turned on the self. How does this happen?

It all has to do with a gene called the *autoimmune regulator* (*AIRE*). The protein from the *AIRE* gene turns on thousands of genes within the thymus. It's like a cellular showroom with parts of cells from your heart, lungs, liver, and brain, all represented within the thymus. Genes that are normally never expressed in the same cell all get turned on.

This process allows for a show-and-tell of cellular parts from all over the body to be expressed within the thymus and then shown to the T cells, which have arrived from the bone marrow to begin their final education. If a T cell's killing apparatus recognizes anything in the body's showroom within the thymus, these T cells are ordered to kill themselves.

This allows for the beta testing of T cells to see if they would potentially react with any part of the body. This complex mechanism is called central tolerance. It means that T cells that are autoreactive to the body are not released from the thymus and into the periphery. This process happens in both men and women, but with one very significant difference. Women do not use their *AIRE* gene to the same degree as men. Why not?

After puberty in women, an estrogen called estradiol downregulates the *AIRE* gene, making it a lot less active. Without an active *AIRE*, fewer of the body's own genes are represented within the showroom of the thymus.

So why are females disproportionately attacking themselves immunologically?

In women, T cells that would normally be instructed to kill themselves because they would invariably attack the body are spared and do not die. This lack of T cell education in women means more of these cells graduate and leave the thymus.

The cost of having a very active immune system leads to a

phenomenon akin to friendly fire: women's bodies sometimes think they are under microbial attack when in fact they're not.

This is what I think of as the Red Riding Hood defense. If a microbe is the wolf and it's dressing up like Grandma, better to kill Grandma every once in a while than to risk being fooled by a wolf dressed like Grandma. One way our bodies employ this strategy is by keeping around T cells, which can target our very own cells and tissues. These T cells are kept just in case there's a pathogen that can trick the immune system, because it appears identical to a native part of the body even though it's not.

Having our immune system employ the Red Riding Hood defense is crucial to combat microbes that are especially adept at changing their outward appearance. It happens every season with different strains of the flu. The influenza virus is constantly shape-shifting to avoid our immunological memories and defenses. So our bodies always need to be on the lookout. That's why even an individual without an autoimmune disease always has some self-directed T cells that are targeting their own body. It's the body's way of making sure that there isn't a new wolf dressed as Grandma lurking somewhere, ready to pounce.

Women do this by releasing a higher percentage of T cells from the thymus that will react and target their own bodies. That makes T cells from genetic females much more likely to attack something that resembles the body—like a hidden wolf—or unfortunately the body itself. Being able to kill wolves dressed like Grandma is a good thing. Occasionally though,

T cells in women unfortunately kill an innocent Grandma instead of a wolf dressed like one. When that happens often, an autoimmune condition like lupus develops.

Hormonal influences might also explain why the rates of autoimmune diseases between the sexes take such a drastic turn after puberty in girls, when the levels of sex hormones like estrogens increase. Although lupus is rare before puberty, the ratio for prepubescent lupus is two females to every one male. After puberty, it jumps to nine females to every one male. This same pattern is seen not only in lupus but in MS as well.

Yet the effects of estrogens are not so straightforward. Estrogens can have different effects based on the amount that's present in the body. When present in lower amounts, estrogens can stimulate the immune system; but in higher concentrations, estrogens can call off or suppress an immune attack.

The reverse happens in males.

After puberty in males, a form of testosterone called dihydrotestosterone increases, causing *AIRE* to become much more active in the thymus. Because *AIRE* is turned on, T cells in genetic males are subjected to a more brutal and rigorous education within the showroom of the thymus than those in females are.

Many more of the T cells that would have recognized the body are told to die. This is good for the wolf, because a wolf dressed up as Grandma would not be attacked by a T cell. That's why T cells in males are even more tolerant (immunologically speaking).

The downside for men is that more wolves dressed like

Grandma escape detection, and this is one reason that men are weaker from an immunological perspective. It's also why they don't get autoimmune conditions to the same degree women do. Their immune system just isn't as critical.

There are other possible causes for the heavy burden that women pay for their genetic superiority. For example, studies have found that female patients with autoimmune diseases have skewed inactivation of one X chromosome. Uneven skewing of one X chromosome over the other within the showroom of the thymus gland would make a woman's T cells more critical of her own body. This would cause a lack of proper T cell education toward the underrepresented X chromosome in the thymus. If there's X skewing in other tissues or organs in the body, a conflict can occur in which T cells recognize them as foreign.

What we don't know yet is if the X inactivation skewing toward one X chromosome, which can happen only in genetic females, is causing autoimmunity or is caused by it. There's also the possibility of genes escaping inactivation on the silenced X, which would cause problems as well. Remember, around 23 percent of the genes on the X that we initially thought were silent aren't. Then there are estrogens that stimulate lymphocytes to secrete cytokines and promote survival of autoreactive T and B cells that then attack their own bodies. When it comes to the differences between the immune system of genetic females and males, a lot of complex mechanisms are at work.

Autoimmunity or autoreactivity is always the major

challenge for the adaptive immune system. But the news isn't completely negative when it comes to autoimmunity for women. Being at higher risk for autoimmune conditions may give females an advantage, and not only when it comes to killing microbes. It may even make female immune cells better cancer killers.

In my opinion, targeting a cancer cell is a form of autoimmunity. Women are more resistant to and better at fighting certain cancers than men, and I would argue that this is an important extension of their immune privilege.

Over millions of years, our genes have come up with creative solutions to get cells to work for the common good of the body. When you need to heal after a cut, cellular growth has to be a tightly controlled process. Many different checks on cellular growth work together to keep all your cells under control. Yet the longer you live, the more likely there will be damage to one of these growth-control mechanisms. Cancer is the result of mutinous cells that grow uncontrollably. This is why cancer is often the inevitable consequence of aging.

Men are more likely to get cancer, and they're also more likely to die from it. According to data compiled by the American Cancer Society, males have a 20 percent higher risk than females of developing cancer, and a 40 percent greater chance of dying from it. The latest numbers of new cases of cancers from the Surveillance, Epidemiology, and End Results (SEER) Program of the National Cancer Institute (NCI) in the United States profoundly illustrate this sex discrepancy. They report that men outnumber women in new diagnoses for the following cancers: bladder, colon and rectum, kidney and renal-pelvis, liver, lung and bronchus, non-Hodgkin's lymphoma, and pancreatic cancer.

The explanation for why men are more susceptible to cancers in general than genetic females goes beyond merely behavioral factors. This is evident when considering that male predominance in developing cancers is also seen in the most common type of leukemia that affects children, acute lymphocytic leukemia (ALL), where boys consistently outnumber girls.* Not all cancers in organs that both sexes share are more common in men. Some cancers, like breast and thyroid, are more commonly diagnosed in women.

But for some types of cancer, such as renal cell carcinoma, two males are diagnosed for every female. This is after adjusting for geographic region, gross domestic product, environmental risk factors, and even tobacco use (differences between men and women used to be greater because more men used tobacco in the past). In the United States, this translates to about 153,000 more new cases of cancer in men every year than in women.

So how do other animals with long life spans, like African and Asian elephants, stay cancer-free? Both the African and the Asian elephant have multiple copies of a gene called *TP53*. When it's working normally, the *TP53* gene is a very important regulator of cellular growth and a crucial stopgap

---

*Not all childhood leukemias predominantly affect males. Less common types of leukemia such as acute myeloid leukemia (AML) occur in males and females about equally.

impeding the multiplication of malignant cells. If you knock out *TP53*, you get unrestricted cellular growth.

If you have two functional copies, like most humans do, then you are two steps away from that happening. If you're an elephant, on the other hand, you have twenty extra copies of *TP53*, which is a lot of backup copies. An elephant trying to fight off cancer has options.

We don't know exactly why or how elephants inherited so many copies of the *TP53* tumor-suppressor gene and whether or not they all work, but it probably has something to do with their incredible size and the amazing number of cells they have—about one hundred times the quantity found in humans.

But one implication of having more cells is cancer. The more cells you have, the greater the possibility that one of them may decide to go rogue. It takes only one cell to bring the whole interconnected system crashing down. That's where having all those extra copies of *TP53* in each and every cell comes in handy. This ensures that cellular order is maintained throughout millions of cells over decades of life.

Elephants also have extra copies of a gene called *LIF*, which is genetic shorthand for leukemia inhibitory factor.* The key word is "inhibitory," and one of those copies of *LIF*, named *LIF6*, maintains fidelity to its name—at the behest of *TP53*, this gene sabotages the cellular machinery of any rogue cell, leading to its death. This is like having a built-in chemotherapy

---

*As with the extra copies of the *TP53* gene found in elephants, it's still not evident whether all the extra copies of *LIF* are fully functional.

cancer-treatment system available on demand—a good thing if you're a very large elephant living a long life. The more we study animals like African and Asian elephants, the more we'll discover about how to better treat cancer in humans.

Now, neither women nor men have multiple pairs of the genes *TP53* or *LIF*. But females do have a way out of the traditional pathway to developing cancer. Women have escape from X-inactivation tumor-suppressor (EXITS) genes,* which refers to the six tumor-suppressor genes that are found on the X chromosome. Instead of the multiple copies of *TP53* or *LIF* that elephants have, all XX females have multiple copies of working EXITS genes.

When these genes are mutated during one's life, there's a significantly higher chance of developing cancer—especially in males. That's because males have only one copy of each of these genes in every one of their cells. Females always have two copies of these tumor-suppressing genes in each cell. Males don't have EXITS genes; females do.† When it comes to cancer prevention, females have options.

When women do develop cancer in organs they share with men, it often develops later on in life and is less aggressive. In some cases, studies have shown that women even respond better to cancer treatment than men and overall have better rates of survival. But the price of being more immune to

---

*Among those genes that escape inactivation in XX females from their inactivated or "silenced" X are the six tumor-suppressor genes: *ATRX*, *CNKSR2*, *DDX3X*, *KDM5C*, *KDM6A*, and *MAGEC3*.
†Males may have similar genes on the Y chromosome, but they do not seem to offer men the same benefits of cancer prevention.

cancer is having a higher rate of almost all other autoimmune diseases.

The first and second line of women's unparalleled cancer defense system is made up of having both X chromosomes cooperating and working together to stop a cell from going rogue, and a much stronger immune response to kill any cells that do.

Like a police state that cracks down hard on any form of citizen dissent, women's cells are not always kind to their own bodies—often inflicting debilitating damage that results in a myriad of autoimmune diseases. Making better antibodies and more aggressive T cells not only helps women fight cancer more effectively but also ensures their survival in overcoming whatever pathogenic obstacles are placed before them.

When considering life and death, survival and extinction, perhaps the cost of superimmunity is worth paying.

6

WELL-BEING: WHY WOMEN'S HEALTH IS NOT MEN'S HEALTH

T**HE PRACTICE OF MEDICINE** was built using research that was done primarily on male cells, male animals, and male test subjects. As a result, we tend to know more about men when it comes to the determinants of health and well-being. With a few exceptions, we clinically treat women just like we treat men.*

---

*Some of these exceptions include gynecological and obstetric issues, and conditions like osteoporosis.

Progress in addressing the differences between the sexes in the practice of clinical medicine has been sluggish. For the most part, this is because the medical establishment has been ignorant of the profound chromosomal uniqueness of genetic females. We didn't understand the fact that female cells can cooperate genetically and that women actually harness the genetic power of their silenced X chromosome in every one of their cells. And of course there's the matter of genetic females' innate immune privilege, which results in women being better suited to fight both infections and cancer. While we now know that this privilege comes at the high price of experiencing more autoimmune diseases, there is no denying the inherent strength and versatility that women enjoy genetically by virtue of their two X chromosomes. And all these crucial differences have been underestimated when it comes to developing, testing, and implementing medical advancements.

I first discovered the depth of this problematic chasm while I was in the early stages of developing my first antibiotic directed to fight against multidrug-resistant superbug microorganisms such as methicillin-resistant *Staphylococcus aureus* (MRSA). Years before researchers actually test a drug or treatment on humans, they are required by government agencies like the Food and Drug Administration (FDA) to go through a pre-clinical stage of research. This often involves using cells and nonhuman animals to produce evidence that the proposed treatment is both efficacious and safe.

When it comes to metals such as zinc and iron, the male and female body have different intake requirements.* Given that some of my antibiotic compounds were metal-based, I wanted to test specifically whether there would be any experimental differences in the results between male and female mice.

As I alluded to in the introduction, the problem was that I couldn't easily get any female mice. I was perplexed when I found out that normally only male mice were used to perform these types of early infectious model experiments.

The FDA did publish a document, all the way back in the year 1987, that provided guidance about the use of both sexes of animals in clinical trials for scientists seeking approval for a new drug or treatment. It stated the following: "Both sexes of animals are to be included in pre-clinical drug safety studies for products targeted for use by both sexes." The only issue was that it was a suggestion rather than a regulation. This particular suggestion didn't have to be followed to get a drug approved by the FDA.

I learned that if I wanted to secure an equal number of male and female mice for my studies, I would actually have to special-order the female mice, as most research-animal breeding facilities didn't readily keep them in stock at the time. Understanding for the first time how unusual it was to request female mice, I realized that the majority of my col-

---

*The recommended dietary allowance (RDA) differs for adult males and females over the age of nineteen years, which for zinc is 11 milligrams and 8 milligrams, respectively. The RDA for iron for adults between the ages of nineteen and fifty years for males is 8 milligrams and for females is 18 milligrams.

leagues were simply using male mice to conduct their preclinical research.

Ordering the female mice would delay the start date of my experiment by many months, throwing my project behind schedule. I wish I had waited. As I was to discover years later when I finally included both female and male mice in my preclinical research studies, the experimental results obtained from only using male mice were different. This differential finding required me to rethink and rework some of my subsequent drug-design strategies. And if that was my experience, maybe the results that other scientists were getting from using only male mice in very early drug discovery accurately predicted clinical outcomes only half the time.

⸻

INCLUDING FEMALE MICE in preclinical research may not entirely solve this problem either. Most of the female mice that are used today in research have been inbred for many generations. That means that unlike human females, who have two vastly different X chromosomes in all their cells, inbred female mice have two identical X chromosomes instead (in effect making them almost like male mice genetically). The implications are that inbred female mice do not have the genetic diversity of human females or female mice that have not been inbred, nor do they benefit from genetic cooperation in the same way. So even if we begin using more female mice in research, we will need to take this important nuance into consideration.

Only relatively recent clinical research has begun to take genetic sex into account. Research in the 1980s and 1990s that was looking at new drug applications (the first step in the long and arduous process of drug approval) found that although women were included in clinical trials, their numbers were still overall underrepresented in many of the studies.

This discrepancy prompted the National Institutes of Health in 1993 to finally require the inclusion of women in NIH-funded clinical research. The latest study to tackle the issue of the inclusion of women in clinical trials looked at about 185,000 clinical trial participants and found no evidence of a significant underrepresentation of women. That's good news—an important step in the right direction. But since most of our prior medical research essentially ignored the differences between the sexes, we still have a lot of refining to do.

Even with the inclusion of women in clinical trials, the sex and gender differences when it comes to drugs and medical procedures haven't yet been fully addressed by all those involved in research. When you look at new drug applications with the FDA, for example, you don't see sex-specific dosage recommendations. This is the case even though these drugs are metabolized and excreted differently by men and women.

Take alcohol, for example. Ethanol is one of the most commonly consumed recreational drugs worldwide. And on average, women metabolize alcohol more slowly than men do. This means that women, with every additional drink imbibed, will suffer more than men from the adverse effects of alcohol consumption.

There are many other examples of differential drug metabolism between the sexes. While training to become a physician, I was taught to prescribe the same dose of the sleeping aid Ambien (zolpidem) to both women and men. Why would I have discriminated between the genetic sexes in my dosing?

It turns out that not discriminating between the genetic sexes in this case can be dangerous. After many years and millions of prescriptions written and filled, some reports began to surface that women were more sensitive than men to the somnolent effects of Ambien. This eventually prompted a safety review of Ambien. No one could have expected what was discovered.

The FDA finally recognized in April 2013 that Ambien requires a different dose for each sex. What most physicians didn't know prior to the announcement was that women metabolize drugs like Ambien more slowly than men. That's why a woman taking the previously recommended dose of Ambien would awake the next morning feeling sluggish and groggy, while a man doing the same would usually awaken rested and refreshed. Consequently, new guidelines from the FDA decreased the dose for women from 10 milligrams to 5 milligrams.*

There is no doubt about it: genetic male and female bodies simply absorb, distribute, metabolize, and eliminate drugs differently. Even over-the-counter medications like Tylenol

---

*Product labeling for Ambien (zolpidem) dosing was halved in 2013 for women, from 10 milligrams to 5 milligrams (immediate-release) daily and from 12.5 to 6.25 milligrams (controlled-release) daily, while staying the same for men.

(acetaminophen), for example, are cleared and removed from the body at a different rate in men—to the tune of 22 percent faster. With all our advancements in studying the human genome since its sequencing in the early part of the twenty-first century, we don't yet understand the underlying genetic pathways that explain this divergence between the genetic sexes.

The specialty that deals with the way the human body handles drugs is called pharmacokinetics. People working in this field have long known that there are significant differences between the sexes. Every one of the above pharmacokinetic factors (such as absorption and elimination) can either increase or decrease the levels of a drug in the body, depending on someone's genetic sex. This means that for one sex a certain dose of a drug can become toxic or poisonous. Alternatively, a drug can be broken down so quickly by one sex that its efficacy is reduced or eliminated completely.

The consequence of determining the preclinical safety and efficacy profile of drugs using only male cells and animals is that women are at greater risk than men to develop an adverse reaction to certain prescription drugs. Clinical drug trials don't always take into account a woman's distinct way of processing drugs because they're designed with knowledge gleaned from preclinical trials that may have used only male cells and male animals. The reason that they don't include female cells and animals is that they're not required to by the agencies that approve drugs, like the FDA in the United States.

As a result, women's divergent abilities when it comes to how they handle certain drugs often don't get tested before the drug is prescribed to them in a clinical trial or after a drug is approved. For example, a heart medication can be life-threatening to a woman if we don't take into account the fact that the rhythm of a woman's heart—the way it beats and moves blood around the body—is sensitive to some prescription drugs. Certain drugs have been pulled from the market because of their increased risk of causing a fatal arrhythmia (for example, torsades de pointes) in women. If early studies and clinical trials had included equal numbers of men and women, perhaps this would have been prevented.

For many years, women had been taking drugs like the antihistamine Seldane (terfenadine) or Propulsid (cisapride) for relief of nighttime heartburn, unaware that with every pill they were at greater risk of disturbing the underlying rhythm of their hearts. We still don't know the extent to which other drugs might influence women's hearts in this destructive way.

We know that it takes genetic females much longer to clear the cardiac drug digoxin, for example, which may be due to a lower activity of hepatic UDP-glucuronosyltransferases in women. These enzymes break down many of the toxic compounds we consume, as well as many of the prescription drugs we take.

In general, females have much longer transverse colons than males. Women also have slower gastric motility and intestinal transit, which means that whatever a woman eats takes longer to make it through and out the other end of the digestive tract. What this means practically for women is that they need to wait for a longer period of time after eating before they take a drug that should be taken on an empty stomach, such as the allergy medication Claritin (loratadine). This strategy ensures that the stomach is sufficiently empty to maximize the absorption of the medication.

Further complicating matters, some drugs work only for women and not for men. Zelnorm (tegaserod) is one such drug. It was approved only for women who suffer from irritable bowel syndrome with constipation, as it was found to be ineffective for men. If Zelnorm had been tested for efficacy before women were participating in clinical research, its benefits for women would have remained undiscovered.

Similarly, at a recent meeting of the NIH Advisory Committee on Research on Women's Health, Dr. Louise McCullough, a stroke researcher, came to the stunning conclusion that she was getting skewed results in her research findings because the male and female cells from the mice she was using had different ischemic death pathways. This means that even though the male and female cells superficially appeared indistinguishable, they were dying in a very distinguishable manner (a surprising finding given that male and female cells were erroneously thought to behave, in dying as in living, in the same manner). McCullough's findings prompted researchers to begin questioning whether perhaps, if these cells were not dying in the same manner, some of their essential cellular life processes were different as well. What she discovered were new avenues of sex-dependent research that could ultimately lead to more effective treatments for both men and women.

Our comprehension of the staggering medical impact

stemming from the differences between the sexes is only in its infancy. As more women begin participating in clinical trials for new drug treatments, and as we reassess medical wisdom from the past through this lens, our knowledge will surely grow.

—

AS I'VE LEARNED OVER THE YEARS, there's still a lot of ground left to be covered when it comes to the differences in human anatomy between the sexes. Until now, most people would have assumed that just about everything there is to know about the anatomy of the human body is known. And that's mostly true—if humanity consisted only of men.

I met Stephanie in my final year of medical school. She was in her midforties and had made an appointment for what she described to me as a long-standing and embarrassing problem that had become more troublesome after she'd had her first child. My role that day was to find out more about Stephanie's medical history, including her current symptoms.

Her primary care physician had referred her to a urologist who specialized in performing urethral-sling procedures. She was referred for surgery because she was experiencing stress urinary incontinence—the medical term for the unintentional loss of urine when the bladder is put under pressure from something like a cough, laugh, or sneeze. Stephanie had some basic questions about the surgery and what she could expect.

I asked her the usual screening questions concerning some of the triggers for urinary leakage that are commonly associated with stress incontinence. She answered no to most of the questions on my prepared checklist. Putting my list and clipboard aside, I asked Stephanie to describe to me exactly what she was experiencing.

"Well . . . it doesn't happen every time. But when it does, it's while my partner and I are having sex. Toward the end, as I'm about to have an orgasm, I feel like I need to pee, and then it happens. When it does, it's just really wet and uncomfortable. My partner is understanding . . . It doesn't seem to bother him, but it bothers me. The worst part is that I can't seem to stop it," she said.

Her symptoms didn't really sound like the symptoms of stress incontinence. Or like anything else that was listed in my intake questionnaire. The uncontrolled loss of urine during sex (or coital incontinence) can happen to women, but it's not usually associated with orgasm, as Stephanie described.

I presented Stephanie's case to the surgeon and relayed to him everything she shared. He thanked me for my intake and let me know that the symptoms for urinary stress incontinence can often present in a myriad of ways.

I found out a few months later that Stephanie did indeed have her surgery, but it wasn't successful. That part wasn't too surprising, as the short-term "cure" rate of this type of procedure is never really 100 percent—somewhere around 80 percent is more realistic. But I still had a nagging concern that something else was going on in Stephanie's case. It turned out that the real "problem" was not incontinence but female ejaculation.

As I can attest from my own clinical training and experience, mainstream medicine in practice is still largely silent on this particular aspect of female anatomy and sexuality. Physicians rarely learn anything of it during their training.

More than fifteen hundred years ago, however, both Aristotle and the physician Galen were well aware of women being able to produce an emission of a "female fluid." Many of their contemporaries also believed that it was the equivalent of male semen and the mixing of both is what resulted in a pregnancy. But where was such a fluid coming from in women?

The female prostate gland of course. And this gland is also not a recent discovery. In the seventeenth century, the Dutch anatomist and physician Regnier de Graaf wrote in great detail about female genital anatomy, which he reported on after his careful anatomical dissections. This included a description of what he called the "female prostatae," which he likened to the male prostate. De Graaf even differentiated between fluid coming from the female prostate and vaginal secretions providing lubrication during coitus.

De Graaf wasn't alone in identifying this female anatomical attribute and ability. The Englishman William Smellie, who was practicing in the eighteenth century, described female ejaculation as a "fluid ejected from the prostate or analogous glands" produced by women.

But current medical thinking dictated that Stephanie was dealing with incontinence. No one even considered any other clinical explanation, and this misunderstanding can be traced back to a nineteenth-century Scottish gynecologist by the name of Alexander Skene. Even today, the clinical anatomy textbooks that professors and medical students rely on worldwide all have one glaring omission—or, should I say, mislabeling.

Skene identified a very small gland with tiny openings on either side of the urethra, with one important oversight. Two hundred years prior to Skene, de Graaf described the exact same gland. De Graaf believed that it released fluid into the urethra directly and, most significantly, that it was the source of female ejaculation. I never learned about any of de Graaf's work, but I did hear about Skene's.

When you look up Skene in a clinical anatomy textbook today, you'll find at least a page or two dedicated to "Skene's glands" (not the female prostate). So, what we are still calling Skene's glands is essentially the female prostate. Skene didn't have a fulsome idea of the function of the gland that he was describing or that the female prostate is related to the male prostate embryologically.

In 2001, the Federative International Committee on Anatomical Terminology officially made the name change from "Skene's glands" to the "female prostate." Not every woman has noticed whether she produces any fluid from her prostate, but one thing's for sure—every genetic female has one. We now also know that fluid released by some women can contain prostate-specific antigen (PSA) and prostate acid phosphatase (PAP), which were once thought to be solely produced by the male prostate.

But strangely, many medical textbooks haven't been corrected or updated, as of this writing, to reflect this name change. That's why some physicians would assume that coital incontinence, the unintended release of urine during sex, could be the only explanation for any extraneous fluid released during female arousal.

In the twenty-first century, Stephanie was treated based on an outdated nineteenth-century model of female anatomy and physiology. Ironically, if modern medicine had been using de Graaf's three-hundred-year-old description of the female prostate, Stephanie's symptoms may not have been pathologized and surgically mistreated.

More recently, my experience with Stephanie's case provided the necessary clue that helped me with another patient with a seemingly unrelated medical condition. Samantha was a healthy forty-one-year-old female who was referred for a consultation from a concierge medical clinic after undergoing an extensive executive physical examination, provided by her new employer, that came up with an unexplained testing result.

Except for the occasional acute migraine headache, for which she took the medication Treximet (sumatriptan and naproxen), Samantha was otherwise healthy. About six months prior to her referral, she had a hormonal IUD placed, without any complications.

The specific reason for Samantha's referral was that a medical error occurred during her executive checkup. Even in the best of health care systems, medical errors still happen and can run the gamut from minor to serious.

When she first came to the concierge clinic, the staff inadvertently registered her as a male, as Samantha often goes by Sam. I knew from personal experience how often this can happen, as my first name, Sharon, is relatively uncommon for a male. Whenever I've gone to see a new physician for a physical, the staff often sets up the room in advance for a gynecological exam, which in my case isn't medically necessary.

Since Samantha was registered as a male in the electronic medical records of the clinic, a panel of male-specific blood tests was reflexively ordered when her blood sample was sent off to the lab for analysis. Surprisingly, Samantha's bloodwork returned with an elevated prostate-specific antigen level. A PSA level below 4.0 nanograms per milliliter is considered by most physicians to be normal. Although there is still some debate today about using PSA levels for prostate cancer screening, a PSA level of 43.2 nanograms per milliliter (which was Samantha's test result) would almost immediately lead to a strong recommendation for imaging and a biopsy of the prostate—if Samantha were male. But Samantha was a genetic female and accordingly was not considered by modern medicine to have a prostate. So why did Samantha have a highly elevated PSA level?

With no real explanation as to how or why she would have an elevated PSA level, she was referred for a more extensive medical workup.

The urologist I referred her to specialized in urologic oncology, and Samantha was found to have developed a Skene's gland adenocarcinoma, a.k.a. prostate cancer. Her elevated

PSA level was in fact because Samantha had a cancer that almost never develops in genetic females. She subsequently underwent surgery, after which her elevated PSA level resolved.

My experience with Stephanie taught me that females have a prostate, and not to assume otherwise. Samantha's case then taught me that although it's extremely rare, having a prostate can result in prostate cancer for women as well. Just as we used to assume that breast cancer affected only females, we have now come to understand that both genetic sexes can be affected by a breast cancer diagnosis.

Medical errors almost never turn out well, but in Samantha's case, the error thankfully did. Because she was initially medically treated as a male, instead of as a female, this particular error very likely ended up saving her life. My hope is that as we expand our knowledge about the differences and similarities between the genetic sexes, we will come to a greater understanding of how to treat them both.

———

BEFORE WE MOVE on to speak more about the future of medicine, we need take a short trip into its recent past, by stepping into the Italian city of Bologna. Ancient porticos and covered walkways have provided shelter and comfort to the pedestrians of Bologna for more than one thousand years. The city itself became a thriving metropolis under the Romans more than two thousand years ago. Today, Bologna is known by the nickname *la dotta*, *la grassa*, *la rossa*, which translates as "the learned," for the ancient university; "the fat," because it's said to be the birthplace of mortadella, ragù, and tortellini; and "the red," for the color of the bricks that are embedded in its towers, walls, and palazzos.

As an increasing number of people moved to the city and the need for more housing arose, with nowhere left to build, extensions to existing houses literally began to spill over onto the street. Eventually this resulted in about twenty-five miles of porticos. Today, these porticos are often filled with the thousands of students who have come from all over Italy to study at the University of Bologna.

Founded in the eleventh century, the university is the oldest continuously operating institution for higher learning in the Western world. Illustrious alumni and teaching faculty have studied at the university, including the inventor of the radio, Guglielmo Marconi, and the Italian poet Durante degli Alighieri, known as Dante.

As I walked down through the porticos of Bologna, the iron-red color of the bricks reminded me of the reason I was in Italy in the first place. I was giving a lecture on my research examining the genetic role that iron plays in the development of human disease.

Specifically, my research focused on what was at the time a little-known genetic condition called hereditary hemochromatosis. Hemochromatosis causes the body to absorb too much iron from the diet. The gene associated with hemochromatosis is called *HFE*, and it's found on chromosome 6.

My work with hemochromatosis began more than twenty

years ago, and at the time, most of medicine believed that the disease I was studying was rare. But despite this perceived rarity, the negative health effects of this particular genetic condition are preventable with an already available treatment. I believed then as I do now in the importance of trying to increase awareness of hemochromatosis to help people who may not know they are affected.

We now know that hemochromatosis is a "silent killer." It's caused by one of the most common mutations in people of western and northern European descent, with up to a third of males having at least one version of the mutant gene, written as *C282Y* or *H63D*. As the iron builds up in the body of those affected, oxidative stress induces a damaging biological process of "rusting." Someone with untreated hemochromatosis becomes like the Tin Man—susceptible to rusting from the inside out. This affects many of the joints of the body and can cause those with hemochromatosis to require hip-replacement surgery. Eventually, organs such as the liver and heart fail as they become too damaged by hemochromatosis to sustain life.

What surprises some people about hemochromatosis is that the genetic mutation isn't linked to the X chromosome, but the condition still affects more males. This is because most genetic females lose iron either through menstruation or pregnancy—life events that naturally reduce the amount of iron within the blood and body. This is why most women are naturally protected from hemochromatosis. For women who do experience this condition, the onset usually occurs after menopause, as excess iron is no longer lost in menstrual blood.

To this day, the treatment for hemochromatosis involves regular sessions of phlebotomy or bleeding, which is similar to (but safer than) the bloodletting commonly practiced centuries ago by cutting veins open with a lancet.

As I walked through the porticos of Bologna, south down Via dell'Archiginnasio, I was distracted by the image of a barber's pole on a poster hung under the porticos. The barber's pole was the original symbol, centuries ago, that advertised the place where bloodletting was done by the barbers or surgeons in the area. It's amazing that doctors today treat people with hemochromatosis using a process that was available along this very street hundreds of years ago.

With all this thinking about blood, I inadvertently missed the entrance to Palazzo dell'Archiginnasio di Bologna and had to turn around. When I arrived there, I admired the gateway to what I perceive as the past and future of medicine. Much of our current understanding of medicine originates from the dissections of the human body that took place here in Bologna and other Italian cities like Padua. This unassuming walkway was the exact same path many of the giants of early medicine took every day on their way to work.

I proceeded to the anatomical theater of the Archiginnasio, which today houses a reproduction of an awe-inspiring room originally built in 1637 where people traveled from all corners of Europe to learn the latest and most advanced medical science of their time. The original anatomical theater was almost completely destroyed by an Allied bomb during an air raid in the waning days of World War II.

In the contemporary reproduction, there's a single slab of marble sitting in the middle of the medieval amphitheater. That's where human bodies were slowly and carefully dissected as curious onlookers sat transfixed by what was happening before them.

Sitting on one of the benches overlooking the anatomical table, I was struck by how little has changed. The dead still have much to teach the living.

It's been years since I left behind the dissection room where I first worked through the secrets of the human body. My anatomical theater didn't have spruce wood paneling, or a quietly imposing seventeenth-century statue of Apollo attached to the ceiling keeping watch over me as I worked. My room had a cream-colored linoleum floor, with a body sitting on a stainless-steel gurney, its head propped up by a well-seasoned wooden block. The view from my room, however, was much more striking: the twenty-first-century Manhattan skyline.

The art of prosection* and the study of human anatomy have not fundamentally changed, because the human body hasn't changed. But the ways in which a human body is procured have changed dramatically. In earlier times, the bodies were not gifts pledged by the living seeking to improve the study of medicine. Many of the bodies dissected

---

*Prosection involves the dissection of a cadaver for the instructive purposes of understanding anatomy.

here in public were stolen or acquired after the work of the executioner's ax or hangman's noose was done. Since many more males were executed, there were more male cadavers available to dissect and closely study.

Although fewer women were executed, they often ended up on the dissection table too, along with women who had died from complications of childbirth. What's clear from all the anatomical study was the fascination at the time regarding the differences between the sexes. Particular attention was paid to the reproductive anatomy of women, especially the uterus.

Detailed and realistic human anatomical models of men, women, babies, and fetuses were also prepared in Italian cities like Florence. These unnerving models, a combination of tissue, bone, and wax, afforded a view into the body uncompromised by the stench of decaying human flesh. Many of these wax models are still on display at the University of Bologna.

One wax model in particular caught my attention while I was visiting—that of a pregnant young woman named Venerina, meaning "little Venus." She was created by Clemente Susini, a famous Florentine anatomical wax artist in the eighteenth century. Venerina is a faithful replica of a young woman who died more than two hundred years ago. If we are to believe that this was an actual representation of a once living person, and the established opinion is that it is, then she likely stood at a height of about four feet nine and died while still pregnant in her teens.

Viewing Venerina is not for the faint of heart. But in her

death, she has something to impart to those who have the patience to carefully watch and learn.

She is lying on her back, safely behind a glass enclosure. Venerina's abdominal and thoracic walls were designed to be removable. This allowed for a virtual dissection, as she can be disassembled, exposing her many different internal organs. The heart was left within her chest but cut open, exposing both the right and left ventricles, or chambers. But if you approach the glass enclosure and look closely, you'll observe something unusual in Venerina's wax heart.

The ventricles or chambers of her heart are of the same thickness, which is not a normal finding in most people. The left ventricle of the heart is usually thicker, as it has to work harder against more pressure while pumping arterial blood. The reason Venerina's ventricles are of equal thickness is present as well—faithfully re-created in wax. There's a small duct that connects her aorta to her pulmonary artery. Today, we call Venerina's case one of a patent ductus arteriosus (PDA), which describes a condition in which a duct that's normally present during fetal life remains open after birth and into adulthood. We now know that an isolated PDA is twice as common in females, and yet we still do not know why. A PDA causes venous and arterial blood to mix and the pressures within the heart to equalize. This explains why the thickness of Venerina's ventricles are abnormal and of equal size. All this from a wax model.

Such a detailed and comprehensive view wasn't available to me during my anatomical dissections more than two hundred years later—without the help of a computer simulation. It's the attention to detail that animates these models, making them feel real. Some of the models are so lifelike, I almost expected them to climb out of their display cases and follow me back to my hotel.

As medicine became modernized, we lost the appreciation for the very subtle and not-so-subtle differences between the sexes that I saw in Bologna. Apprehending these differences carries life-and-death consequences for the practice of medicine. Often this involves visually inspecting a patient for the telltale signs of a disease, like those signs in Venerina.

Studies have found much variability in a physician's ability to make a visual diagnosis. Sometimes you just need to know where to look and what to look for. Take malignant melanoma. The best chance a patient has for survival is still through an early and careful visual diagnosis. Malignant melanoma is the least common type of skin cancer, but that doesn't stop it from being the deadliest—especially for older white males. Now that should be fairly predictable given the fact that the lighter the skin pigmentation, the more likely someone is to suffer the ill effects of DNA-damaging ultraviolet radiation from the sun. That still doesn't explain why males would suffer from higher rates of melanoma than women.

Staying out of the sun is still one of the best preventative measures against melanoma. Back when I was studying potatoes in the Altiplano in Peru, where UV radiation is 30 percent higher than it is at sea level, I tried to remember to diligently apply sunscreen, always wear a hat, and stay out of the sun during the peak hours of 11:00 a.m. to 2:00 p.m. Unfortunately, all those things didn't happen very often. This was not good news, considering that genetic males are both more prone to getting melanoma and more likely to have a worse prognosis than any woman—and a much lower cure rate to boot.

Research indicates that melanoma is predicated on not only genetic biology but behavior as well. This is why the location of melanoma is different between the sexes—on the back and trunk for males and the lower legs for females. People generally follow the dictates of the fashion of their time, which can mean dressing in clothes that expose certain parts of the body more than others. This is probably why there are variations in the location of melanoma on the body for genetic males and females. UV radiation from sun exposure, after all, is the most significant environmental risk factor for melanoma.

There are many behavioral influences, so it's difficult to tease out the exact reason for the difference in melanoma rates and outcomes between the sexes. What we do know is that genetic females' immune privilege is likely one of the significant reasons that women have an advantage in keeping skin cancer at bay and overcoming it as well.

Melanoma isn't the only cancer for which we observe sex differences in prevalence and treatment outcomes. As mentioned previously, colorectal cancer is more common in males. When women do develop colorectal cancer, it's often found in their right colon; while in men, it's usually found in the left colon. We don't know the reason for this, but the difference does have real-world implications. The polyps in the colons of genetic females that eventually develop into cancers grow higher up and are more likely to be missed during sigmoidos-copies. In addition, a woman's diagnosis of colorectal cancer typically happens five years later than a man's. This is why continuing with colonoscopy screening later on in life might be more beneficial for women than men in detecting right-sided colon cancers.

Another difference between the sexes that we're still in the process of discovering is the development of lung cancer in both men and women who do not smoke. For reasons yet unknown, female nonsmokers are more likely to develop lung cancer than male nonsmokers are. Genetic men who smoke seem to be predisposed to developing lung cancer at a higher rate than women smokers.

Suffice it to say that wherever we look in the human body, we are finding that female and male organs don't behave in the same way. Such differences between the sexes shouldn't be surprising when we consider that every one of our cells has a sex, and therefore the tissues, organs, and bodies that are made up of these cells have a genetic sex as well.

▭

MEDICINE RARELY PAYS much attention to how the same type of injury can affect the sexes disproportionately. We're start-ing to realize that traumatic brain injuries (TBIs) are one such example of this. A TBI is caused by a blow, jolt, or shock to the head that disrupts or changes the functioning of the brain as a result. A TBI can also be caused by a sudden acceleration or deceleration of the head, which jostles the softer jellylike brain around within the skull. Sudden forceful movements can also initiate shearing forces, which shatter the delicate architecture of the brain.

But not all TBIs are the same. They can happen as a result of a range of injuries, from a minor concussion, which is considered a mild TBI, to a life-threatening TBI that requires immediate medical intervention.

This is what happened to one of the boxing world's longest and oldest reigning champions, the Canadian Adonis Stevenson, who suffered a severe TBI in a match defending his title in December 2018. During the eleventh round of a fight with Oleksandr Gvozdyk, Stevenson received a flurry of mighty blows to the head that knocked him out flat. He struggled to get up and stumbled. Something was obviously wrong. If it weren't for emergency surgery and intensive follow-up care, most doctors agree that Stevenson would surely have died.

Not everyone who experiences a TBI knows they have suffered one at the time. Sometimes the effects of a TBI take years to manifest, and they can spark personality changes as well as changes in the way the brain functions. Other times, the effects of TBI are so obvious that even someone watching on live television can readily observe their severity.

Most people who suffer from TBIs today are men. So, we

are just beginning to appreciate the long-term effects of TBIs in women, and the results are concerning. Studies looking at sports with similar rules for both women and men—like basketball and soccer—have found that women not only sustained more concussions than males but also reported worse long-term symptoms as a result of them.

In addition, the physical proportions of the neck and head between the sexes are, on average, different: women experience greater angular acceleration of the head when hit—all resulting in much more serious TBIs.

My patient Lorena taught me firsthand the role that a TBI can play in changing a person's life course. When I first met Lorena, she was lying on her back in a bright orange jumpsuit, her hands handcuffed to a hospital bed. She glanced at me disdainfully as I entered her room. "Fuck off" was all she said.

I took a deep breath. I was in my fourth year of medical training at the time, completing a final round of internal medicine at an inner-city hospital in New York. The senior medical staff on call had assigned me to Lorena's case, and I was responsible for her care. Her room was dimly lit, and as I got closer to Lorena, I saw that her face seemed rather pale and she looked exhausted. I noticed that there was a glint of light coming off something metallic near her exposed feet. I then understood that her legs were shackled together.

I hadn't treated anyone within the correctional system before my work with Lorena and wasn't sure what her medical history entailed before she arrived at the hospital on a frigid

February morning with two armed guards in tow. According to her medical chart, she was brought to the hospital because she had fainted twice in the last two weeks.

Her medical records indicated that she was experiencing what was assumed to be a very heavy menstrual flow. It had started four weeks ago and hadn't yet abated. This obviously wasn't normal. I tried to talk with Lorena about her symptoms, but apart from her initial sunny greeting, she didn't say much else. Given her medical presentation, I was more than a little concerned.

I finished writing up my notes based on our initial meeting and put in a request for bloodwork, some basic imaging, and a consultation from the gynecology staff to try to clarify why Lorena was bleeding so heavily.

The only other significant issue in Lorena's medical history was a severe TBI she sustained during a lacrosse match in high school. After her TBI, her family and friends reported marked personality changes and personal difficulties. But since the TBI occurred when she was an adolescent, I didn't consider it germane to her initial presentation to the emergency department.

The explanation for Lorena's bleeding came the following day. My pager went off and I returned the call. It was the gynecology resident, who said that Lorena's continuous bleeding was likely due to a very advanced gynecological malignancy, probably a stage IV cervical cancer. The resident couldn't confirm her clinical suspicions, though, because Lorena refused to be biopsied or examined any further.

While I was still on the phone, my pager went off again. I recognized the extension—it was an urgent page from the hospital lab. I wrapped up my call and dialed the number. A laboratory tech picked up, and she was curt and straight to the point: "Patient number XX from this morning's blood labs: hemoglobin level critical at five g/dL." The line clicked. Lorena was in trouble.

A hemoglobin level is an indication of the body's capacity to move life-sustaining oxygen from the outside world into the cells that need it. If you cut off or impair that ability below a certain point, you suffocate from the inside out, one cell at a time. This is one of the rare exceptions when the general practice of medicine actually takes genetic sex into account. For a woman, any hemoglobin level below 12 grams per deciliter is associated with anemia. For men, the level is a little higher, at 13 grams per deciliter. A level below 6 or 7 grams per deciliter for either sex usually requires a swift blood transfusion.

This was on my mind as I rode the elevator up to Lorena's floor. Without her permission, there was no way to give her the blood transfusion she most definitely needed. And without that transfusion, her life would be in danger.

I heard from one of the nurses who was caring for Lorena that she liked Diet Coke, so I stopped by a vending machine on my way to her room. Maybe a peace offering would help. Thankfully, after discussing the importance of getting the blood transfusion and later the biopsy, Lorena consented to both. I left her room with my mood elevated by the one small victory of knowing that at least now with the transfusion she'd be out of imminent danger.

After arranging for the transfusion, I moved on to the other medical responsibilities I had for my remaining patients that day. An hour later my pager went off. It was a call from Lorena's nurse. "She's now refused the transfusion and threatened my staff. We're calling it off." I asked the nurse if Lorena was still bleeding. She replied in the affirmative.

"That's not a good sign. Let me speak to her again . . . Maybe I can change her mind. I'm on my way up," I said.

Given that she was still bleeding, it was likely that Lorena's hemoglobin was dropping even lower. I explained again the medical reasons why she needed help and the risks that she was taking by refusing it. Lorena relented and consented for the transfusion again.

Disaster averted. Or so I thought. Only a half hour later my pager went off again. It was Lorena's nurse. "She's refusing the transfusion again. Do you want to come back to speak to her?"

I returned to her room and reiterated my concerns. Lorena seemed convinced and agreed to the transfusion for the third time, only to refuse it once more minutes after I left her room.

I returned to her bedside. "Lorena, this can't go on. You're continuing to lose blood, and without a transfusion you are seriously putting your life at risk," I said. "My shift is about to end and I want to leave knowing you're out of danger. The other staff is waiting outside of your room with a bag of blood that can save your life. Once it leaves the blood bank, it can't go back for safety reasons. And because you're O negative, you can only receive O negative blood. O negative blood is rare, so if you're thinking about refusing it again, keep in mind that this pint of rare blood could have helped someone else."

She remained quiet as she processed what I told her and then said in an even and quiet tone, "Fine. I'll do it for real this time." And with that, I arranged for the transfusion.

It never happened.

I was leaving the hospital when I heard the code blue announced over the PA system. I ran back to Lorena's room. The crash cart was already out and the code leader was calling out orders. A member of the code team was compressing her chest while another doctor was trying desperately to insert an IV.

Lorena was pronounced dead shortly thereafter.

We now know unequivocally that physical trauma affects the brain in a myriad of complex and lifelong ways. The most likely explanation for the personality changes Lorena experienced was that they were a direct consequence of the trauma to her brain. We still don't know enough about TBIs, and when it comes to the effects of TBIs on the brains of women, we know even less. Many of the changes to the brain following an injury cause permanent alterations in how it works, leading to a deficit in executive functions, which ultimately affects a person's emotional, cognitive, and social functioning.

Mounting evidence is pointing to a very serious phenomenon: all other things being equal, when the same physical force is applied to the brain, it's experienced differently between the sexes. Even though the studies have been small so far, they are indicating that women are at greater risk for sustaining a sports-related TBI than men, as well as experiencing worse outcomes.

We will not be fully aware of the differences that result from similar injuries between women and men until we start looking. An illustrated example of this can be seen from the results of a recent study examining the structural, metabolic, and functional brain alterations that can result from sports-related repetitive subconcussive head impacts. Athletes may not even be aware that they have suffered a subconcussive head impact during the normal course of play.

In the study, twenty-five collegiate-level ice-hockey players (fourteen male and eleven female) underwent diffusion-weighted magnetic resonance imaging (dMRI) before and after their hockey season. The stunning maps of the kaleidoscopic colors of the white matter tracks that are produced from dMRI are so visually striking that they have made their way into art galleries.

The imaging produced by dMRI provides a valuable way of assessing the state of the wiring of the brain—an area that has been reported in the past to be quite sensitive to damage, especially from shearing forces experienced from physical trauma. Suffice it to say, damage to the internal wiring of the brain can have dire consequences.

The researchers' study revealed that significant changes were seen on brain imaging at the end of the hockey season in the superior longitudinal fasciculus, internal capsule, and corona radiata of the right hemisphere of the brain. This is consistent with the type of damage seen in people who have experienced a TBI.

Yet none of the twenty-five players in the study subjectively reported experiencing any type of head injury whatsoever, though the dMRI showed otherwise. And not only that, but none of the brain changes that were observed through the dMRI happened to the male hockey players. It was only the female hockey players who were found to have experienced brain changes at the end of the hockey season. No doubt as we advance in our medical and neurological understanding of how TBIs affect women's brains specifically, we'll do a far more efficacious job of treating them.

PHYSICIANS CAN LEARN the most important and enduring lessons from our patients. For me, Amanda was another one of those patients, and her case was a crash course in the deeper limitations of modern medicine when it comes to treating women.

In the field of cardiovascular disease, it is established scientific fact that there are significant differences between the sexes. Yet these differences are still often overlooked in the latest practice recommendations. It hasn't even been that long since we made the elementary realization that women and men present with different symptoms when they are experiencing a myocardial infarction, or heart attack. It is against this medical backdrop that I first met Amanda.

I was handed Amanda's case, a holdover from a late Saturday night, just as I started my shift early Sunday morning at a very busy New York City hospital.

At forty-seven, Amanda was the picture of perfect health. She exercised almost daily and made sure to eat a balanced diet with a big focus on fresh fruits and vegetables. Obesity was common in Amanda's family, as was non-insulin-dependent diabetes, and the thought of developing either is what kept her motivated, especially on those days after work when she felt like skipping the gym and grabbing a few cocktails with her friends instead.

She'd eventually catch up with her friends after exercising and always managed to stay busy and social. As she shared all this with me during my initial examination, she also quietly mentioned that her marriage had recently ended and that she was having a hard time dealing with the ensuing emotional strain.

Thankfully, she wasn't having any thoughts of harming herself, but she was really devastated by what had happened to her. That was understandable, as Amanda found out not only that her husband of twelve years was having an affair with one of her best friends but also that they were now expecting a child together. Her husband came clean only a week before she appeared in the emergency department and was asking for a quick divorce. Given what Amanda was going through at that moment, I thought that she was coping rather well.

The only clue that the medical staff left for me in her chart was a neat three-pronged trident. This was often used as a shorthand symbol for psychiatry. It didn't seem that anyone was taking her medical concerns seriously—staff simply thought that she might need to speak to someone from psychiatry about her recent breakup, which wasn't unreasonable.

When I first saw Amanda seated calmly on a gurney, I observed that she didn't look acutely ill or injured, like most of the other people in the emergency department that morning, which was populated with an array of folks who suffered drunken falls, injuries from fights, and drug overdoses from opioid use.

Amanda's symptoms were vague and nonspecific and consisted mainly of feeling anxious, lethargic, nauseated, and a little soreness in her chest that she attributed to overdoing it at the gym the day before. The physician's assistant who first saw her when she came in thought she might be pregnant and ordered a blood test, a fair assumption given the fact that she missed her last two periods.

Her routine bloodwork and urine sample appeared normal, and her pregnancy test was negative. We ended up discharging Amanda with a follow-up appointment to be seen by someone in the psychiatry department later that same week.

I finished my shift that day and returned the following morning, surprised to see Amanda in the emergency department again, just as she was registering. This time her symptoms had changed to include acute chest pain that seemed to radiate down both her arms. She thought she was having a heart attack. It's hard to believe that we had all missed what was actually wrong with Amanda.

We quickly performed an EKG and sent her blood to the lab to test for the common markers of a heart attack and performed a bedside echocardiogram. Her EKG was, in fact, abnormal, as was her echocardiogram. But actually, Amanda was not having a heart attack. As was immediately evident from her cardiac imaging, Amanda had a ballooning of the left ventricle of her heart, which is associated with a condition called takotsubo cardiomyopathy.

More than 90 percent of people diagnosed with takotsubo cardiomyopathy are women. The condition is named after a Japanese trap designed to capture octopi because it resembles the abnormal shape of the heart that's seen in those with this condition. Even more mysteriously, takotsubo cardiomyopathy always seems to follow an extreme emotional event. Maybe that's why it's also known as "broken-heart syndrome."

Amanda was fortunate and made a full and complete recovery. Once thought to be very rare, takotsubo may be much more common than we've realized, according to recent research. What's interesting is that although it is so much more common in females, when males are affected by takotsubo, they don't recover as well. The likely reason is that, as with the cells that make up our body, our organs also have a very specific sex—one that is chosen long before we are born.

———

EVERY HUMAN KIDNEY is either a male or a female. The kidney is around four to five inches in length and shaped like a giant bean. Each kidney has around one million nephrons, which help the body filter the blood. Every time the heart beats, blood makes its way to the kidneys for filtration through the nephrons. They reabsorb what the body wants to keep and filter out or excrete toxic waste products from everyday living, the result of which is the production of urine. When protein consumption increases, the kidneys have to work much harder to get rid of all the waste products created from specifically metabolizing protein. That's one of the reasons why people suffering from chronic kidney disease and waiting for a transplant are often told to be judicious when it comes to their protein consumption.

Around one hundred thousand people are currently waiting for a new kidney in the United States today. The majority of them will die waiting. Every ten minutes there's someone added to the list of those in need of a transplant, be it for a liver, a heart, or lungs. Another twenty people who are already listed for a transplant will die waiting for a new organ every day in the United States alone.

Kidneys fail for many reasons. An autoimmune condition (like lupus nephritis, which cost Selena Gomez the use of her kidneys) is one way this happens. High blood pressure, diabetes, and blockages in the blood vessels to the kidney, called renal artery stenosis, are other reasons people end up requiring a new kidney through surgical transplantation.

Overall, male kidneys contain more nephrons, while female kidneys have 10 to 15 percent less. That means a male kidney has a greater overall capability of filtering the blood than a female kidney does. Given the choice, it's better to receive a kidney with more horsepower than not.

If both kidneys fail, the only way to survive while waiting for a new kidney is by undergoing dialysis. This is the artificial process of trying to filter the waste out of the blood. It's nowhere near as efficient as what one's natural kidneys can achieve. The best outcome for those who receive a transplanted kidney results from obtaining one from a living donor. Most of the people on dialysis attest to the profound life changes that happen after receiving their new kidney. It's immediate and consequential. Although it's not a permanent solution for most people, it is the only thing that can save and prolong their life.

The majority of the people in need of a new kidney are genetic males. And most of the living donors who are providing them are genetic females. The reason more men need new kidneys may have to do with sex-related biological factors like hypertension (high blood pressure), which men suffer from in greater numbers, which then leads to kidney damage.

Several clinical trials have found that receiving an organ donated by a genetic female is a risk factor for both rejection of that organ and death for the men who receive it. In addition, when it comes to a woman in need of a kidney transplant, research indicates that the success rate for transplants is higher when she receives a male kidney, as opposed to a female one. Other studies have found that men who received other female organs, like a heart or a liver, had the worst outcomes.

Why is that?

One reason may have to do with the fact that, as I mentioned earlier, all organs have a sex (male or female) because they are made up of cells that have one. Male kidney cells are less sensitive to many of the side effects of the immunosuppressive drugs that people need to take to stop their body from attacking the "foreign" organ. This means that when a female patient receives a male kidney, her new male kidney may experience fewer side effects from the drugs she'll be administered during her recovery process.

The other important reason is that all the cells within male organs, like the kidney, are using the exact same X chromosome. In the female kidney, there are a combination of cells using different X chromosomes. This makes a female kidney far more genetically diverse and therefore more immunogenic than a male kidney. This can help explain why more female organs than male organs are rejected by the body of the recipient.

Some of these differences in outcome may also have to do with the quality of the organs that are being donated. Most of the female donors are older than the male ones, while the reverse is true in the case of recipients. Many males who are receiving a transplant might also be in poorer health. Ultimately, the most distinguishing factor is the organs' sex.

As we've seen, there are many differences between men and women when it comes to health outcomes, longevity, and disability. Many of these differences are related to the sex of the cells, tissues, and organs that support the life of the body.

To truly advance the medical science behind women's health, we need to include more women in research and also find better ways to effectively compare research results between women and men. For instance, we know that women are more susceptible to ischemic strokes as well as Alzheimer's disease.

As medical research has only recently begun to consider the differences between the genetic sexes, we don't yet have a good theoretical understanding and explanation as to why more women are currently being diagnosed with conditions like Alzheimer's disease. We still don't know exactly how to do a better job at helping these women. This is why researchers will need to consciously and consistently use female cells and animals in their studies. Understanding the differences between women and men will allow us to better help both.

Simply continuing to argue for more research in and of itself is not enough when that very knowledge generation is predicated on a conceptual framework that is male-based. What we need is a completely novel gaze when it comes to medical research and practice as it pertains to women. That's why we must first take into account females' unique genetic diversity and cellular cooperation, which results in their overall genetic superiority.

And we need to promptly apply this new understanding

when treating and researching female conditions. Medically, this means that we should not devise research studies that look at females solely through a biological male lens.

As clinicians, researchers, and the public at large begin to grasp the genetic chasm dividing the chromosomal sexes, medicine will have to work hard to begin applying this knowledge to practice. The important work is only just beginning.

CONCLUSION:
WHY SEX CHROMOSOMES MATTER

MANY OF US DON'T GIVE MUCH THOUGHT to the sex chromosomes we've inherited. Yet these microscopic strands of DNA play a fundamental role in every aspect of our lives, though most of us haven't even had the opportunity to get up close and actually see them. If your sex chromosomes have been functioning properly, silently and diligently, why would you need to think about them? After all, they've been working since long before you were born: the X chromosome you received from your mother was created while she was still in her own mother's womb, and so on. If you received a Y chromosome, then you got it from your father, who got it from his.

The first time I extracted my chromosomes from within the leukocyte cells in my blood, I was struck by how small the Y chromosome was. While preparing my karyotype, which is a process of visualizing and identifying each of my forty-six chromosomes by size and unique banding patterns, the Y chromosome was immediately the easiest to recognize. As I arranged my chromosomes in their respective pairs, the Y chromosome, minuscule and alone, was left without a mate. When I finally placed my Y chromosome next to my X chromosome, it was in that moment that I first visually comprehended how much more genetic material each female actually possesses.

Yet, throughout college, graduate school, and even in medical school, I heard an incessant buzz about the importance of the Y chromosome to the human species. After all, as I was told, it's what makes a man. There are a lot of reasons for this focus, but I think it also must have had something to do with the fact that most of the people who were speaking breathlessly about the Y had one as well.

Of our twenty-three chromosomes, the one I almost never heard anything about—except in negative terms—was the X chromosome. There were endless lectures about all the various problems that the X caused—everything from colorblindness to intellectual disabilities. When a chromosome was thought to be misbehaving, there was always one that was brought out in front of the whole class and scolded like an errant child: the

X chromosome. Not much has changed since then, as most of medical research and the practice of medicine today has continued to study the X chromosome in terms of its negative health implications.

And yet as you now know, this is all perfectly correct. That is, if you're a genetic male. However, if you were born with two X chromosomes, then instead of being colorblind, you might be a tetrachromat, able to see millions more colors than a genetic male sees. And instead of having a damaged immune system, you have a *stronger* one, allowing you to tackle the most serious infections that would flatten the average genetic male.

So, yes, the differences that are the result of inherited sex chromosomes are profound. The reason your doctors may not be aware of how important it is to consider your sex chromosomes when prescribing medication or screening you for cancer is not due to any willing ignorance on their part. For many years, there has been a serious lack of inclusion of females in all levels of medical research, which has trickled down to how medicine is taught and, often, how it is practiced. Thankfully, that's beginning to change.

My firsthand scientific experience has taught me much of what I know today about females' inherent genetic advantage over genetic males. But it's always my personal experiences that take this knowledge from the theoretical to the painfully practical.

While preparing for our honeymoon, which would have us spend a few weeks exploring the ruins of the Angkor Wat temple complexes in Cambodia, my future wife, Emma, and I both received vaccinations for typhoid fever.

The bacteria *Salmonella typhi* can cause a dreadful infection that's often acquired from less-than-hygienic food-preparation practices. Far from benign, typhoid fever can kill a fifth of the people infected if they go untreated.

Although we both got the exact same vaccination at exactly the same time, I returned to work the next day, while Emma did not. Based on the differences between our physical reaction to the vaccine, it was as if we were given two different vaccinations. Her arm was so sore at the injection site that I had to help her dress, and a headache with malaise kept her in bed for the remainder of the week.

When I spoke with the travel-medicine nurse who vaccinated us, she told me that in her experience what was happening to us was common—women just seem to react more vigorously to the shots. I pretty much felt nothing after my vaccine. And for this I was to pay dearly.

Three months later, we found ourselves exploring the steamy jungles of Cambodia when I noticed that I was feeling a little off. At first, I thought it might be jet lag or maybe the oppressive heat, but soon I found myself in the midst of a full-blown episode of typhoid fever.

While I was in bed in the hospital, I remember looking up and staring at the IV bag, which contained a lifesaving course of antibiotics, and thinking that my wife and I had shared the same meals since we arrived in Cambodia. So why was I sick with a food-borne pathogen while she sat by my hospital bed, perfectly fine? As I came to learn, she didn't suffer needlessly after her vaccination for typhoid after all.

While my immune system seemed to ignore the vaccine, hers didn't, gearing her body up for what we were both about to face in Cambodia. By responding with both X chromosomes to the vaccine, my partner's immune system was reacting to the vaccine by design and preparing for the worst—an encounter with a real life-threatening microbe. Through a process triggered by somatic hypermutation, her B cells were developing the best-fitting antibodies to target and kill typhoid fever.

Obviously, our bodies didn't respond similarly to the vaccination, even though both of us have the exact same class of immune cells with similar DNA within them, circulating in our blood. While male and female immune cells may share similar DNA, that doesn't mean they are using the same genes to the same degree. Many of the genes related to immunity in my cells stayed quiet following vaccination, while my wife's immune cells and the genes within them responded to the vaccination with the utmost urgency.

Even in genes that we both share that are not immune related, there are actual differences in how each sex uses them. Recent research has discovered that genetic females and males use as many as 6,500 of their 20,000 genes differently across 45 tissues that are common to both sexes. Some of the genes that were found to be more active in men are related to the growth of body hair and building muscle, while other genes involved in fat storage and drug metabolism* were more active in women.

My wife's immunological vigor as exemplified by our contrasting vaccination experience is just one example of how a genetic woman physiologically outplays a genetic male. This strength is also what puts genetic females at a greater lifelong risk of turning and directing their potent immunological defense mechanisms against themselves, which can result in an autoimmune disease.

Ultimately, there's only one way to judge overall superiority between the genetic sexes. The real test of one's mettle is being able to survive the challenges of life. So, who is left standing at the far end of life?

Let's review the numbers: Genetic males begin life with a generous head start when it comes to demography, as on average 105 boys are born for every 100 girls. But as we saw with Jordan and Emily in the NICU at the start of this book, when life begins progressing, that same head start quickly diminishes, eventually disappearing altogether. Around the time people reach the age of 40, the number of females and males becomes about equal. But by the age of 100, about 80 percent

---

*Of the 6,500 genes found to be behaving differently between the sexes, two genes that are involved in drug metabolism, *CYP3A4* and *CYP2B6* (part of the cytochrome P450 family of enzymes), were found to be more active in genetic females. These two genes encode the instructions for two enzymes involved in the metabolism of more than 50 percent of the drugs prescribed today to both males and females. One reason genetic females suffer so many more side effects induced from prescription medications than males do is likely that their genes are behaving and metabolizing drugs differently.

of people alive are female. And 95 percent of supercentenarians (people over the age of 110) are female.

Of the fifteen leading causes of death in the United States, men are overrepresented in thirteen of them. These include a litany of illnesses such as heart disease, cancer, liver disease, kidney disease, and diabetes. Of these top fifteen causes, it's only in Alzheimer's disease that females are overrepresented, and when it comes to cerebrovascular disease, the genetic sexes are tied.

Women outliving men is not an isolated phenomenon specific to the United States. Recently, life expectancy at birth was analyzed around the world and found to be higher in females over males in every one of the fifty-four countries studied.

If outwitting death is the ultimate indicator of genetic strength, then women's resounding success of crossing the supercentenarian finish line makes them the indisputable victors. It should come as no surprise that it is almost always a woman who holds the title of the oldest living person. Kane Tanaka is one of the latest in an incredibly long line of women to hold this distinguished title.

Tanaka's life history is an illustration of the female survival advantage. Even though she was born prematurely on January 2, 1903—the same year the Wright brothers first took to flight—Tanaka ultimately went on to outlive both her husband and her son.

Tanaka described how she'd yearned for the coveted title of the oldest person alive since her 100th birthday. At

116 years of age, Tanaka finally got her wish. Overcome with emotion, she said that being recognized by Guinness World Records as both the oldest living person and the oldest living woman was the most exciting moment of her life. During the ceremony marking the occasion, she was asked pointedly what she'd enjoyed the most in her life, and Tanaka answered, "This right now."

Recent research studying adult sex ratios in 344 species confirmed what I had observed firsthand as a neurogenetics researcher twenty years earlier. More females were alive later in life in those species, like humans, that employ the XY-male and XX-female chromosome system. The opposite pattern was found to occur in species like birds, where the males have the same two sex chromosomes, ZZ, and the females have ZW. Clearly, it's not just the human species that benefits from having and using two identical sex chromosomes.

As I've illustrated in *The Better Half*, the genetic advantage that women possess results from every cell within a female having the option of using one of her two X chromosomes, each of which contains around one thousand genes. The genes that are found on the X chromosome are essential to life, playing a critical role in the development and maintenance of the brain as well as the immune system. As we saw in X-linked genetic conditions, such as X-linked intellectual disabilities and even colorblindness, having that spare X chromosome is invaluable. The genetic diversity and the cellular cooperation that occur within female bodies give them a genetic advantage over the males in our species.

Despite all this, genetic males are not the disposable sex. Obviously, we need both genetic sexes to reproduce and thrive. But it is females who have evolved to be the better half, genetically speaking. The sooner we come to terms with this fact and adjust the way we research and practice medicine, the better for us all.

NOTES

Epigraph

vii *"I have undertaken—as boldly as I can"*: Agrippa, Henricus C. (2007). *Declamation on the Nobility and Preeminence of the Female Sex*. Edited and translated by Albert Rabil. Chicago: University of Chicago Press.

Introduction

3 *Here are some basic facts: Women live longer than men*: If you'd like to read more about sex discrepancy in human longevity, see Ostan R, Monti D, Gueresi P, Bussolotto M, Franceschi C, Baggio G. (2016). Gender, aging and longevity in humans: An update of an intriguing/neglected scenario paving the way to a gender-specific medicine. *Clin Sci (Lond)* 130(19): 1711–1725; Zarulli V, Barthold Jones JA, Oksuzyan A,

Lindahl-Jacobsen R, Christensen K, Vaupel JW. (2018). Women live longer than men even during severe famines and epidemics. *Proc Natl Acad Sci USA* 115(4): E832–E840.

3 *Women have stronger immune systems*: For more information on the variety of differences in the immunological responses between the sexes, see Giefing-Kröll C, Berger P, Lepperdinger G, Grubeck-Loebenstein B. (2015). How sex and age affect immune responses, susceptibility to infections, and response to vaccination. *Aging Cell* 14(3): 309–321; Spolarics Z, Peña G, Qin Y, Donnelly RJ, Livingston DH. (2017). Inherent X-linked genetic variability and cellular mosaicism unique to females contribute to sex-related differences in the innate immune response. *Front Immunol* 8: 1455.

3 *Women are less likely to suffer from a developmental disability*: For an introduction into the male intellectual disability burden, see the following: Muthusamy B, Selvan LDN, Nguyen TT, Manoj J, Stawiski EW, Jaiswal BS, Wang W, Raja R, Ramprasad VL, Gupta R, Murugan S, Kadandale JS, Prasad TSK, Reddy K, Peterson A, Pandey A, Seshagiri S, Girimaji SC, Gowda H. (2017). Next-generation sequencing reveals novel mutations in X-linked intellectual disability. *OMICS* 21(5): 295–303; Niranjan TS, Skinner C, May M, Turner T, Rose R, Stevenson R, Schwartz CE, Wang T. (2015). Affected kindred analysis of human X chromosome exomes to identify novel X-linked intellectual disability genes. *PLoS One* 10(2): e0116454.

3 *are more likely to see the world in a wider variety of colors*: If you'd like to read more about color perception abilities in humans in general, see the following: John D. Mollon, Joel Pokorny, Ken Knoblauch. (2003). *Normal and Defective Colour Vision*. Oxford, UK: Oxford University Press; Kassia St. Claire. (2017). *The Secret Lives of Color*. New York: Penguin; Veronique Greenwood. The humans with super human vision. *Discover*, June 2012; Jameson KA, Highnote SM, Wasserman LM. (2001). Richer color experience in observers with multiple photopigment opsin genes. *Psychon Bull Rev* 8(2): 244–261; Jordan G, Deeb SS, Bosten JM, Mollon JD. (2010). The dimensionality of color vision in carriers of anomalous trichromacy. *J Vis* 10(8): 12.

6 *Elderly women on average outlive*: There's a plethora of published literature on female longevity. If you'd like to read more about this topic, these few articles are a good place to start: Marais GAB, Gaillard JM, Vieira C, Plotton I, Sanlaville D, Gueyffier F, Lemaitre JF. (2018). Sex gap in aging and longevity: Can sex chromosomes play a role? *Biol Sex Differ* 9(1): 33; Pipoly I, Bokony V, Kirkpatrick M, Donald PF, Szekely T, Liker

A. (2015). The genetic sex-determination system predicts adult sex ratios in tetrapods. *Nature* 527(7576): 91–94; Austad SN, Fischer KE. (2016). Sex differences in lifespan. *Cell Metab* 23(6): 1022–1033.

8 *It's no wonder research has shown*: Parra J, de Suremain A, Berne Audeoud F, Ego A, Debillon T. (2017). Sound levels in a neonatal intensive care unit significantly exceeded recommendations, especially inside incubators. *Acta Paediatr* 106(12): 1909–1914; Laubach V, Wilhelm P, Carter K. (2014). Shhh . . . I'm growing: Noise in the NICU. *Nurs Clin North Am* 49(3): 329–344; Almadhoob A, Ohlsson A. (2015). Sound reduction management in the neonatal intensive care unit for preterm or very low birth weight infants. *Cochrane Database Syst Rev* 1: CD010333.

8 *Often one of the biggest challenges for the youngest preemies*: We have made a lot of progress in treating the youngest of patients. For more on this topic, see the following articles: Benavides A, Metzger A, Tereshchenko A, Conrad A, Bell EF, Spencer J, Ross-Sheehy S, Georgieff M, Magnotta V, Nopoulos P. (2019). Sex-specific alterations in preterm brain. *Pediatr Res* 85(1): 55–62; Glass HC, Costarino AT, Stayer SA, Brett CM, Cladis F, Davis PJ. (2015). Outcomes for extremely premature infants. *Anesth Analg* 120(6): 1337–1351; EXPRESS Group, Fellman V, Hellström-Westas L, Norman M, Westgren M, Källén K, Lagercrantz H, Marsál K, Serenius F, Wennergren M. (2009). One-year survival of extremely preterm infants after active perinatal care in Sweden. *JAMA* 301(21): 2225–2233.

9 *thankfully, over time, we've developed improved*: Macho P. (2017). Individualized developmental care in the NICU: A concept analysis. *Adv Neonatal Care* 17(3): 162–174; Doede M, Trinkoff AM, Gurses AP. (2018). Neonatal intensive care unit layout and nurses' work. *HERD* 11(1): 101–118; Stoll BJ, Hansen NI, Bell EF, Walsh MC, Carlo WA, Shankaran S, Laptook AR, Sánchez PJ, Van Meurs KP, Wyckoff M, Das A, Hale EC, Ball MB, Newman NS, Schibler K, Poindexter BB, Kennedy KA, Cotten CM, Watterberg KL, D'Angio CT, DeMauro SB, Truog WE, Devaskar U, Higgins RD; Eunice Kennedy Shriver National Institute of Child Health and Human Development Neonatal Research Network. (2015). Trends in care practices, morbidity, and mortality of extremely preterm neonates, 1993–2012. *JAMA* 314(10): 1039–1051; Stensvold HJ, Klingenberg C, Stoen R, Moster D, Braekke K, Guthe HJ, Astrup H, Rettedal S, Gronn M, Ronnestad AE; Norwegian Neonatal Network. (2017). Neonatal morbidity and 1-year survival of extremely preterm infants. *Pediatrics* 139(3): pii, e20161821.

13 *I knew from my clinical work and research*: The following meta-analysis of nineteen studies that examined the outcomes among trauma patients, looking at 100,566 men and 39,762 women, found that male sex was associated with increased risk of mortality, hospital length of stay, and higher incidence of complications. For more information, see Liu T, Xie J, Yang F, Chen JJ, Li ZF, Yi CL, Gao W, Bai XJ. (2015). The influence of sex on outcomes in trauma patients: A meta-analysis. *Am J Surg* 210(5): 911–921. For further reading about this topic, see the following book and articles: Al-Tarrah K, Moiemen N, Lord JM. (2017). The influence of sex steroid hormones on the response to trauma and burn injury. *Burns Trauma* 5: 29; Bösch F, Angele MK, Chaudry IH. (2018). Gender differences in trauma, shock and sepsis. *Mil Med Res* 5(1): 35; Barbara R. Migeon. (2013). *Females Are Mosaics: X-Inactivation and Sex Differences in Disease*. New York: Oxford University Press; Pape M, Giannakópoulos GF, Zuidema WP, de Lange-Klerk ESM, Toor EJ, Edwards MJR, Verhofstad MHJ, Tromp TN, van Lieshout EMM, Bloemers FW, Geeraedts LMG. (2019). Is there an association between female gender and outcome in severe trauma? A multi-center analysis in the Netherlands. *Scand J Trauma Resusc Emerg Med* 27(1): 16.

13 *This was because her body had the use of two X chromosomes*: Spolarics Z, Peña G, Qin Y, Donnelly RJ, Livingston DH. (2017). Inherent X-linked genetic variability and cellular mosaicism unique to females contribute to sex-related differences in the innate immune response. *Front Immunol* 8: 1455.

14 *The immune systems of genetic females are much more likely*: Billi AC, Kahlenberg JM, Gudjonsson JE. (2019). Sex bias in autoimmunity. *Curr Opin Rheumatol* 31(1): 53–61; Chiaroni-Clarke RC, Munro JE, Ellis JA. (2016). Sex bias in paediatric autoimmune disease—not just about sex hormones? *J Autoimmun* 69: 12–23.

15 *What began as a fifty-fifty split between cells*: Peña G, Michalski C, Donnelly RJ, Qin Y, Sifri ZC, Mosenthal AC, Livingston DH, Spolarics Z. (2017). Trauma-induced acute X chromosome skewing in white blood cells represents an immuno-modulatory mechanism unique to females and a likely contributor to sex-based outcome differences. *Shock* 47(4): 402–408; Chandra R, Federici S, Németh ZH, Csóka B, Thomas JA, Donnelly R, Spolarics Z. (2014). Cellular mosaicism for X-linked polymorphisms and IRAK1 expression presents a distinct phenotype and improves survival following sepsis. *J Leukoc Biol* 95(3): 497–507.

17 *A U.S. General Accounting Office review*: Petkovic J, Trawin J, Dewidar O, Yoganathan M, Tugwell P, Welch V. (2018). Sex/gender reporting

and analysis in Campbell and Cochrane systematic reviews: A cross-sectional methods study. *Syst Rev* 7(1): 113; Sandberg K, Verbalis JG. (2013). Sex and the basic scientist: Is it time to embrace Title IX? *Biol Sex Differ* 4(1): 13.

17 *"being male or female is an important fundamental variable"*: Institute of Medicine (U.S.), Committee on Understanding the Biology of Sex and Gender Differences, Mary-Lou Pardue, Theresa M. Wizemann. (2001). *Exploring the Biological Contributions to Human Health: Does Sex Matter?* Washington, DC: National Academies Press.

1. The Facts of Life

19 *Here's some basic biology that will be required as I develop my argument*: Steven L. Gersen, Martha B. Keagle. (2013). *The Principles of Clinical Cytogenetics*. New York: Humana Press; R. J. McKinlay Gardner, Grant R. Sutherland, Lisa G. Shaffer (2013). *Chromosome Abnormalities and Genetic Counseling*. New York: Oxford University Press; Reed E. Pyeritz, Bruce R. Korf, Wayne W. Grody, eds. (2018). *Emery and Rimoin's Principles and Practice of Medical Genetics and Genomics*. London: Academic Press.

21 *We are well on our way to mastering the arts*: Crawford GE, Ledger WL. (2019). In vitro fertilisation/intracytoplasmic sperm injection beyond 2020. *BJOG* 126(2): 237–243; Vogel G, Enserink M. (2010). Nobel Prizes honor for test tube baby pioneer. *Science* 330(6001): 158–159.

21 *Your biological sex isn't always the same as your gender*: Lina Gálvez, Bernard Harris. (2016). *Gender and Well-Being in Europe: Historical and Contemporary Perspectives*. Abingdon: Routledge; McCauley E. (2017). Challenges in educating patients and parents about differences in sex development. *Am J Med Genet C Semin Med Genet* 175(2): 293–299.

23 *Rarely, a baby born with two X chromosomes can develop*: I wrote about this case in my book *Inheritance: How Our Genes Change Our Lives—and Our Lives Change Our Genes*, published by Grand Central Publishing in 2014. For more information about the *SOX3* gene and its role in XX sex reversal, see the following papers: Moalem S, Babul-Hirji R, Stavropolous DJ, Wherrett D, Bägli DJ, Thomas P, Chitayat D. (2012). XX male sex reversal with genital abnormalities associated with a de novo *SOX3* gene duplication. *Am J Med Genet A* 158A(7): 1759–1764; Vetro A, Dehghani MR, Kraoua L, Giorda R, Beri S, Cardarelli L, Merico M, Manolakos E, Parada-Bustamante A, Castro A, Radi O, Camerino G, Brusco A, Sabaghian M, Sofocleous C, Forzano F, Palumbo P, Palumbo O, Calvano S, Zelante L, Grammatico P, Giglio S, Basly M,

Chaabouni M, Carella M, Russo G, Bonaglia MC, Zuffardi O. (2015). Testis development in the absence of SRY: Chromosomal rearrangements at *SOX9* and *SOX3*. *Eur J Hum Genet* 23(8): 1025–1032; Xia XY, Zhang C, Li TF, Wu QY, Li N, Li WW, Cui YX, Li XJ, Shi YC. (2015). A duplication upstream of *SOX9* was not positively correlated with the SRY-negative 46,XX testicular disorder of sex development: A case report and literature review. *Mol Med Rep* 12(4): 5659–5664.

24 *Historically, the importance of having a male child*: Bhatia R. (2018). *Gender Before Birth: Sex Selection in a Transnational Context.* Seattle: University of Washington Press.

24 *Aristotle documented his findings*: Vergara MN, Canto-Soler MV. (2012). Rediscovering the chick embryo as a model to study retinal development. *Neural Dev* 7: 22; Haqq CM, Donahoe PK. (1998). Regulation of sexual dimorphism in mammals. *Physiol Rev* 78(1): 1–33.

26 *Unfortunately, pregnancy and many of these herbal medicines*: Sex-selection drugs and herbal medicines are estimated to cause thousands of stillbirths as well as increase the risk for congenital malformations. A recent study reported a threefold increased risk when women were consuming herbal medicines during their pregnancy. Here are a few articles and editorials describing the current situation as it stands: Neogi SB, Negandhi PH, Sandhu N, Gupta RK, Ganguli A, Zodpey S, Singh A, Singh A, Gupta R. (2015). Indigenous medicine use for sex selection during pregnancy and risk of congenital malformations: A population-based case-control study in Haryana, India. *Drug Saf* 38(9): 789–797; Neogi SB, Negandhi PH, Ganguli A, Chopra S, Sandhu N, Gupta RK, Zodpey S, Singh A, Singh A, Gupta R. (2015). Consumption of indigenous medicines by pregnant women in North India for selecting sex of the foetus: What can it lead to? *BMC Pregnancy Childbirth* 15: 208.

27 *One of them was Nettie Stevens*: Brush, S. (1978). Nettie M. Stevens and the discovery of sex determination by chromosomes. *Isis* 69(2): 163–172; Wessel GM. (2011). Y does it work this way? Nettie Maria Stevens (July 7, 1861–May 4, 1912). *Mol Reprod Dev* 78(9): Fmi; Ogilvie MB, Choquette CJ. (1981). Nettie Maria Stevens (1861–1912): Her life and contributions to cytogenetics. *Proc Am Philos Soc* 125(4): 292–311.

28 *Another female scientist who doesn't always get her due*: Kalantry S, Mueller JL. (2015). Mary Lyon: A tribute. *Am J Hum Genet* 97(4): 507–510; Rastan S. (2015). Mary F. Lyon (1925–2014). *Nature* 518(7537): 36; Watts G. (2015). Mary Frances Lyon. *Lancet* 385(9970): 768; Morey C, Avner

P. (2011). The demoiselle of X-inactivation: 50 years old and as trendy and mesmerising as ever. *PLoS Genet* 7(7): e1002212.

29 *At this very early stage of pregnancy*: Sahakyan A, Yang Y, Plath K. (2018). The role of Xist in X-chromosome dosage compensation. *Trends Cell Biol* 28(12): 999–1013; Gendrel AV, Heard E. (2014). Noncoding RNAs and epigenetic mechanisms during X-chromosome inactivation. *Annu Rev Cell Dev Biol* 30: 561–580; Wutz A. (2011). Gene silencing in X-chromosome inactivation: Advances in understanding facultative heterochromatin formation. *Nat Rev Genet* 12(8): 542–553.

31 *For most of the fifty years since Lyon's paper about X inactivation*: If you'd like to read Dr. Mary Lyon's original groundbreaking paper, see Lyon, MF. (1961). Gene action in the X-chromosome of the mouse (*Mus musculus L.*). *Nature* 190: 372–373.

33 *As a geneticist, I was fascinated by honey bee reproduction*: Breed MD, Guzmán-Novoa E, Hunt GJ. (2004). Defensive behavior of honey bees: Organization, genetics, and comparisons with other bees. *Annu Rev Entomol* 49: 271–298; Metz BN, Tarpy DR. (2019). Reproductive senescence in drones of the honey bee (*Apis mellifera*). *Insects* 10(1).

35 *Female honey bees have also been discovered to be the advanced mathematicians of the insect world*: Howard SR, Avarguès-Weber A, Garcia JE, Greentree AD, Dyer AG. (2019). Numerical cognition in honeybees enables addition and subtraction. *Sci Adv* 5(2): eaav0961; Howard SR, Avarguès-Weber A, Garcia JE, Greentree AD, Dyer AG. (2019). Symbolic representation of numerosity by honeybees (*Apis mellifera*): Matching characters to small quantities. *Proc Biol Sci* 286(1904): 20190238.

37 *Despite the genetic superiority of females, fewer girls*: The sex-ratio-at-birth data is available for almost all countries worldwide and is kept by UNdata, http://data.un.org/Data.aspx?d=PopDiv&f=variableID%3A52.

2. Resilience: Why Women Are Harder to Break

39 *Dr. Barry J. Marshall was getting desperate*: Enserink M. (2005). Physiology or medicine: Triumph of the ulcer-bug theory. *Science* 310(5745): 34–35; Sobel RK. (2001). Barry Marshall. A gutsy gulp changes medical science. *US News World Rep* 131(7): 59; Kyle RA, Steensma DP, Shampo MA. (2016). Barry James Marshall—discovery of *Helicobacter pylori* as a cause of peptic ulcer. *Mayo Clin Proc* 91(5): e67–68.

40 *"I preferred to believe my eyes, not the medical textbooks"*: Barry J. Marshall, ed. (2002). *Helicobacter Pioneers: Firsthand Accounts from the Scientists Who Discovered Helicobacters, 1892–1982*. Carlton South: Wiley-Blackwell.

41 *"I had this discovery that could undermine a $3 billion industry"*: Pamela Weintraub. The doctor who drank infectious broth, gave himself an ulcer, and solved a medical mystery. *Discover*, April 2010; Groh EM, Hyun N, Check D, Heller T, Ripley RT, Hernandez JM, Graubard BI, Davis JL. (2018). Trends in major gastrectomy for cancer: Frequency and outcomes. *J Gastrointest Surg*. doi: 10.1007/s11605-018-4061-x.

43 *We've known for a very long while that ulcers*: Rosenstock SJ, Jørgensen T. (1995). Prevalence and incidence of peptic ulcer disease in a Danish County—a prospective cohort study. *Gut* 36(6): 819–824; Räihä I, Kemppainen H, Kaprio J, Koskenvuo M, Sourander L. (1998). Lifestyle, stress, and genes in peptic ulcer disease: A nationwide twin cohort study. *Arch Intern Med* 158(7): 698–704.

43 *But now, there is no doubt that it's because males*: Schurz H, Salie M, Tromp G, Hoal EG, Kinnear CJ, Möller M. (2019). The X chromosome and sex-specific effects in infectious disease susceptibility. *Hum Genomics* 13(1): 2; Sakiani S, Olsen NJ, Kovacs WJ. (2013). Gonadal steroids and humoral immunity. *Nat Rev Endocrinol* 9(1): 56–62; Spolarics Z, Peña G, Qin Y, Donnelly RJ, Livingston DH. (2017). Inherent X-linked genetic variability and cellular mosaicism unique to females contribute to sex-related differences in the innate immune response. *Front Immunol* 8: 1455; Ding SZ, Goldberg JB, Hatakeyama M. (2010). *Helicobacter pylori* infection, oncogenic pathways and epigenetic mechanisms in gastric carcinogenesis. *Future Oncol* 6(5): 851–862.

43 *In humans, treating cell lines of human gastric adenocarcinoma*: Ohtani M, Ge Z, García A, Rogers AB, Muthupalani S, Taylor NS, Xu S, Watanabe K, Feng Y, Marini RP, Whary MT, Wang TC, Fox JG. (2011). 17 β-estradiol suppresses *Helicobacter pylori*–induced gastric pathology in male hypergastrinemic INS-GAS mice. *Carcinogenesis* 32(8): 1244–1250; Camargo MC, Goto Y, Zabaleta J, Morgan DR, Correa P, Rabkin CS. (2012). Sex hormones, hormonal interventions, and gastric cancer risk: A meta-analysis. *Cancer Epidemiol Biomarkers Prev* 21(1): 20–38.

44 *Whether it is bacteria such as* Staphylococcus aureus: Schurz H, Salie M, Tromp G, Hoal EG, Kinnear CJ, Möller M. (2019). The X chromosome and sex-specific effects in infectious disease susceptibility. *Hum Genomics* 13(1): 2.

47 *Recently, the Thai government has made immense strides*: To read more about this impressive accomplishment, see the following articles: Lolekha R, Boonsuk S, Plipat T, Martin M, Tonputsa C, Punsuwan N, Naiwatanakul T, Chokephaibulkit K, Thaisri H, Phanuphak P,

Chaivooth S, Ongwandee S, Baipluthong B, Pengjuntr W, Mekton S. (2016). Elimination of mother-to-child transmission of HIV-Thailand. *MMWR Morb Mortal Wkly Rep* 65(22): 562–566; Thisyakorn U. (2017). Elimination of mother-to-child transmission of HIV: Lessons learned from success in Thailand. *Paediatr Int Child Health* 37(2): 99–108.

48 *Today we know that even when they're treated with the same cocktail*: Griesbeck M, Scully E, Altfeld M. (2016). Sex and gender differences in HIV-1 infection. *Clin Sci (Lond)* 130(16): 1435–1451; Jiang H, Yin J, Fan Y, Liu J, Zhang Z, Liu L, Nie S. (2015). Gender difference in advanced HIV disease and late presentation according to European consensus definitions. *Sci Rep* 5: 14543.

49 *Yet, just one year after starting HAART*: Beckham SW, Beyrer C, Luckow P, Doherty M, Negussie EK, Baral SD. (2016). Marked sex differences in all-cause mortality on antiretroviral therapy in low- and middle-income countries: A systematic review and meta-analysis. *J Int AIDS Soc* 19(1): 21106; Kumarasamy N, Venkatesh KK, Cecelia AJ, Devaleenol B, Saghayam S, Yepthomi T, Balakrishnan P, Flanigan T, Solomon S, Mayer KH. (2008). Gender-based differences in treatment and outcome among HIV patients in South India. *J Womens Health* 17(9): 1471–1475.

50 *Genetic females are uniquely evolved to make better-fitting antibodies*: Hwang JK, Alt FW, Yeap LS. (2015). Related mechanisms of antibody somatic hypermutation and class switch recombination. *Microbiol Spectr* 3(1): MDNA3–0037–2014; Kitaura K, Yamashita H, Ayabe H, Shini T, Matsutani T, Suzuki R. (2017). Different somatic hypermutation levels among antibody subclasses disclosed by a new next-generation sequencing-based antibody repertoire analysis. *Front Immunol* 8: 389; Methot SP, Di Noia JM. (2017). Molecular mechanisms of somatic hypermutation and class switch recombination. *Adv Immunol* 33: 37–87; Sheppard EC, Morrish RB, Dillon MJ, Leyland R, Chahwan R. (2018). Epigenomic modifications mediating antibody maturation. *Front Immunol* 9: 355; Xu Z, Pone EJ, Al-Qahtani A, Park SR, Zan H, Casali P. (2007). Regulation of *AICDA* expression and AID activity: Relevance to somatic hypermutation and class switch DNA recombination. *Crit Rev Immunol* 27(4): 367–397; Methot SP, Litzler LC, Subramani PG, Eranki AK, Fifield H, Patenaude AM, Gilmore JC, Santiago GE, Bagci H, Côté JF, Larijani M, Verdun RE, Di Noia JM. (2018). A licensing step links AID to transcription elongation for mutagenesis in B cells. *Nat Commun* 9(1): 1248.

51 *Women have two different X chromosomes in each of their immune cells*: Schurz H, Salie M, Tromp G, Hoal EG, Kinnear CJ, Möller M. (2019). The X chromosome and sex-specific effects in infectious disease susceptibility. *Hum Genomics* 13(1): 2; Spolarics Z, Peña G, Qin Y, Donnelly RJ, Livingston DH. (2017). Inherent X-linked genetic variability and cellular mosaicism unique to females contribute to sex-related differences in the innate immune response. *Front Immunol* 8: 1455; Vázquez-Martínez ER, García-Gómez E, Camacho-Arroyo I, González-Pedrajo B. (2018). Sexual dimorphism in bacterial infections. *Biol Sex Differ* 9(1): 27.

52 *Studies show that breastfed babies even have a reduced risk*: Tromp I, Kiefte-de Jong J, Raat H, Jaddoe V, Franco O, Hofman A, de Jongste J, Moll H. (2017). Breastfeeding and the risk of respiratory tract infections after infancy: The Generation R Study. *PLoS One* 12(2): e0172763; Gerhart KD, Stern DA, Guerra S, Morgan WJ, Martinez FD, Wright AL. (2018). Protective effect of breastfeeding on recurrent cough in adulthood. *Thorax* 73(9): 833–839.

52 H. pylori *can hijack the process of hypermutation*: Ding SZ, Goldberg JB, Hatakeyama M. (2010). *Helicobacter pylori* infection, oncogenic pathways and epigenetic mechanisms in gastric carcinogenesis. *Future Oncol* 6(5): 851–862; Matsumoto Y, Marusawa H, Kinoshita K, Endo Y, Kou T, Morisawa T, Azuma T, Okazaki IM, Honjo T, Chiba T. (2007). *Helicobacter pylori* infection triggers aberrant expression of activation-induced cytidine deaminase in gastric epithelium. *Nat Med* 13(4): 470–476.

54 *Kafka was forty years old on June 3, 1924*: If you would like to learn more about Kafka's personal medical history, see Felisati D, Sperati G. (2005). Famous figures: Franz Kafka (1883–1924). *Acta Otorhino-laryngol Ital* 25(5): 328–332; Mydlík M, Derzsiová K. (2007). Robert Klopstock and Franz Kafka—the friends from Tatranské Matliare (the High Tatras). *Prague Med Rep* 108(2): 191–195; Vilaplana C. (2017). A literary approach to tuberculosis: Lessons learned from Anton Chekhov, Franz Kafka, and Katherine Mansfield. *Int J Infect Dis* 56: 283–285.

55 *an unfortunate incident known as the Lübeck disaster*: Lange L, Pescatore H. (1935). Bakteriologische Untersuchungen zur Lübecker Säuglingstuberkulose. *Arbeiten a d Reichsges-Amt* 69: 205–305; Schuermann P, Kleinschmidt H. (1935). Pathologie und Klinik der Lübecker Säuglingstuberkuloseerkrankungen. *Arbeiten a d Reichsges-Amt* 69: 25–204.

56 *Multidrug-resistant tuberculosis (MDR-TB)*: The World Health Organi-

zation has a repository of comprehensive information regarding tuberculosis, including the number of cases of MDR-TB, which currently stands at approximately 558,000 worldwide. For more information specifically about tuberculosis, the World Health Organization website is a good place to start: https://www.who.int/tb/en/.

57 *Surviving in the pathogenic soup we live in*: If you'd like to read more about the impact of infectious disease and how it has shaped human history in a myriad of ways, see one of my previous books: Sharon Moalem with Jonathan M. Prince. (2007). *Survival of the Sickest: A Medical Maverick Discovers Why We Need Disease*. New York: William Morrow.

3. Disadvantaged: The Male Brain

60 *For a long time, the commonly held belief was that a boy*: Loomes R, Hull L, Mandy WPL. (2017). What is the male-to-female ratio in autism spectrum disorder? A systematic review and meta-analysis. *J Am Acad Child Adolesc Psychiatry* 56(6): 466–474.

60 *According to the 2018 numbers published*: Kogan MD, Vladutiu CJ, Schieve LA, Ghandour RM, Blumberg SJ, Zablotsky B, Perrin JM, Shattuck P, Kuhlthau KA, Harwood RL, Lu MC. (2018). The prevalence of parent-reported autism spectrum disorder among US children. *Pediatrics* 142(6); Christensen DL, Braun KVN, Baio J, Bilder D, Charles J, Constantino JN, Daniels J, Durkin MS, Fitzgerald RT, Kurzius-Spencer M, Lee LC, Pettygrove S, Robinson C, Schulz E, Wells C, Wingate MS, Zahorodny W, Yeargin-Allsopp M. (2018). Prevalence and characteristics of autism spectrum disorder among children aged 8 years—Autism and Developmental Disabilities Monitoring Network, 11 Sites, United States, 2012. *MMWR Surveill Summ* 65(13): 1–23.

61 *The lack of X chromosomal variety within the cells*: Benavides A, Metzger A, Tereshchenko A, Conrad A, Bell EF, Spencer J, Ross-Sheehy S, Georgieff M, Magnotta V, Nopoulos P. (2019). Sex-specific alterations in preterm brain. *Pediatr Res* 85(1): 55–62; Skiöld B, Alexandrou G, Padilla N, Blennow M, Vollmer B, Adén U. (2014). Sex differences in outcome and associations with neonatal brain morphology in extremely preterm children. *J Pediatr* 164(5): 1012–1018; Zhou L, Zhao Y, Liu X, Kuang W, Zhu H, Dai J, He M, Lui S, Kemp GJ, Gong Q. (2018). Brain gray and white matter abnormalities in preterm-born adolescents: A meta-analysis of voxel-based morphometry studies. *PLoS One* 13(10): e0203498; Hintz SR, Kendrick DE, Vohr BR, Kenneth Poole W, Higgins RD; NICHD Neonatal Research Network. (2006).

Gender differences in neurodevelopmental outcomes among extremely preterm, extremely-low-birthweight infants. *Acta Paediatr* 95(10): 1239–1248.

61 *Of the one thousand genes on the X chromosome, more than one hundred*: Neri G, Schwartz CE, Lubs HA, Stevenson RE. (2017). X-linked intellectual disability update. *Am J Med Genet A* 176(6): 1375–1388; Takashi Sado. (2018). *X-Chromosome Inactivation: Methods and Protocols*. New York: Springer Nature; Stevenson RE, Schwartz CE. (2009). X-linked intellectual disability: Unique vulnerability of the male genome. *Dev Disabil Res Rev* 15(4): 361–368.

62 *Symptoms of X-linked intellectual disability*: Lubs HA, Stevenson RE, Schwartz CE. (2012). Fragile X and X-linked intellectual disability: Four decades of discovery. *Am J Hum Genet* 90(4): 579–590; Roger E. Stevenson, Charles E. Schwartz, R. Curtis Rogers. (2012). *Atlas of X-Linked Intellectual Disability Syndromes*. New York: Oxford University Press.

62 *Almost 99 percent of those affected with fragile X*: Hagerman RJ, Berry-Kravis E, Hazlett HC, Bailey DB Jr, Moine H, Kooy RF, Tassone F, Gantois I, Sonenberg N, Mandel JL, Hagerman PJ. (2012). Fragile X syndrome. *Nat Rev Dis Primers* 3: 17065; Bagni C, Tassone F, Neri G, Hagerman R. (2012). Fragile X syndrome: Causes, diagnosis, mechanisms, and therapeutics. *J Clin Invest* 122(12): 4314–4322; Bagni C, Oostra BA. (2013). Fragile X syndrome: From protein function to therapy. *Am J Med Genet A* 161A(11): 2809–2821; Lubs HA, Stevenson RE, Schwartz CE. (2012). Fragile X and X-linked intellectual disability: Four decades of discovery. *Am J Hum Genet* 90(4): 579–590.

64 *We now know that boys are at a disadvantage*: Boyle CA, Boulet S, Schieve LA, Cohen RA, Blumberg SJ, Yeargin-Allsopp M, Visser S, Kogan MD. (2011). Trends in the prevalence of developmental disabilities in US children, 1997–2008. *Pediatrics* 127(6): 1034–1042; Xu G, Strathearn L, Liu B, Yang B, Bao W. (2018). Twenty-year trends in diagnosed attention-deficit/hyperactivity disorder among US children and adolescents, 1997–2016. *JAMA Netw Open* 1(4): e181471.

64 *An impressive study carried out in Finland*: Gissler M, Järvelin MR, Louhiala P, Hemminki E. (1999). Boys have more health problems in childhood than girls: Follow-up of the 1987 Finnish birth cohort. *Acta Paediatr* 88(3): 310–314.

64 *In a significant study published by the CDC in 2011*: Boyle CA, Boulet S, Schieve LA, Cohen RA, Blumberg SJ, Yeargin-Allsopp M, Visser S, Kogan MD. (2011). Trends in the prevalence of developmental disabil-

ities in US children, 1997–2008. *Pediatrics* 127(6): 1034–1042. See the Centers for Disease Control and Prevention's website for more information: https://www.cdc.gov/ncbddd/developmentaldisabilities/features /birthdefects-dd-keyfindings.html.

64 *The latest numbers published for the United States*: The following publication discusses the prevalence of developmental disability diagnosed in children ages three through seventeen years in the United States for 2014–2016. The result was reported as 8.15 percent for males and 4.29 percent for females. For more information, see the following: Zablotsky B, Black LI, Blumberg SJ. (2017). Estimated prevalence of children with diagnosed developmental disabilities in the United States, 2014–2016. *NCHS Data Brief* 291: 1–8.

65 *The brain is not a simple organ*: For a greater understanding about the myriad of complex developmental processes involved in brain development, see my book *Inheritance: How Our Genes Change Our Lives—and Our Lives Change Our Genes*, published by Grand Central Publishing in 2014.

65 *When babies have trouble eating or sticking out their tongues*: Hong P. (2013). Five things to know about . . . ankyloglossia (tongue-tie). *CMAJ* 185(2): E128; Power RF, Murphy JF. (2015). Tongue-tie and frenotomy in infants with breastfeeding difficulties: Achieving a balance. *Arch Dis Child* 100(5): 489–494.

66 *Clubfoot, or talipes—a condition*: Congenital talipes equinovarus is a common orthopedic foot deformity in children. The Ponseti method, which uses casts, is one of the preferred methods for treating talipes today. See the following articles for more information on the condition and a review of different treatment modalities: Ganesan B, Luximon A, Al-Jumaily A, Balasankar SK, Naik GR. (2017). Ponseti method in the management of clubfoot under 2 years of age: A systematic review. *PLoS One* 12(6): e0178299; Michalski AM, Richardson SD, Browne ML, Carmichael SL, Canfield MA, Van Zutphen AR, Anderka MT, Marshall EG, Druschel CM. (2015). Sex ratios among infants with birth defects, National Birth Defects Prevention Study, 1997–2009. *Am J Med Genet A* 167A(5): 1071–1081.

67 *At their genetic best, men can only aspire to have normal color vision*: John D. Mollon, Joel Pokorny, Ken Knoblauch. (2003). *Normal and Defective Colour Vision*. Oxford, UK: Oxford University Press.

67 *Having the use of two X chromosomes with different versions*: Neitz J, Neitz M. (2011). The genetics of normal and defective color vision. *Vision Res* 51(7): 633–651; Simunovic MP. (2010). Colour vision deficiency. *Eye(Lond)* 24(5): 747–755.

67 *This supercharged version of the color world*: The following article is thought to contain the first reference to the possibility of tetrachromacy or tetrachromatic vision in humans: de Vries H. (1948). The fundamental response curves of normal and abnormal dichromatic and trichromatic eyes. *Physica* 14(6): 367–380. For more detailed information on trichromatic and tetrachromatic vision, see the following articles: Jordan G, Deeb SS, Bosten JM, Mollon JD. (2010). The dimensionality of color vision in carriers of anomalous trichromacy. *J Vis* 10(8): 12; Jameson KA, Highnote SM, Wasserman LM. (2001). Richer color experience in observers with multiple photopigment opsin genes. *Psychon Bull Rev* 8(2): 244–261; Kawamura S. (2016). Color vision diversity and significance in primates inferred from genetic and field studies. *Genes Genomics* 38: 779–791; Neitz J, Neitz M. (2011). The genetics of normal and defective color vision. *Vision Res* 51(7): 633–651; Veronique Greenwood. The humans with super human vision. *Discover*, June 2012.

68 *The incredible thing about our eyes*: Lamb TD. (2016). Why rods and cones? *Eye (Lond)* 30(2): 179–185; Lamb TD, Collin SP, Pugh EN Jr. (2007). Evolution of the vertebrate eye: Opsins, photoreceptors, retina and eye cup. *Nat Rev Neurosci* 8(12): 960–976; Nickle B, Robinson PR. (2007). The opsins of the vertebrate retina: Insights from structural, biochemical, and evolutionary studies. *Cell Mol Life Sci* 64(22): 2917–2932.

68 *Each type of cone cell uses a receptor*: Kassia St. Claire. (2017). *The Secret Lives of Color*. New York: Penguin; Xie JZ, Tarczy-Hornoch K, Lin J, Cotter SA, Torres M, Varma R; Multi-Ethnic Pediatric Eye Disease Study Group. (2014). Color vision deficiency in preschool children: The multi-ethnic pediatric eye disease study. *Ophthalmology* (7): 1469–1474; Yokoyama S, Xing J, Liu Y, Faggionato D, Altun A, Starmer WT. (2014). Epistatic adaptive evolution of human color vision. *PLoS Genet* 10(12): e1004884.

69 *Scientists found that the colorblind male monkeys*: Troscianko J, Wilson-Aggarwal J, Griffiths D, Spottiswoode CN, Stevens M. (2017). Relative advantages of dichromatic and trichromatic color vision in camouflage breaking. *Behav Ecol* 28(2): 556–564; Doron R, Sterkin A, Fried M, Yehezkel O, Lev M, Belkin M, Rosner M, Solomon AS, Mandel Y, Polat U. (2019). Spatial visual function in anomalous trichromats: Is less more? *PLoS One* 14(1): e0209662; Melin AD, Chiou KL, Walco ER, Bergstrom ML, Kawamura S, Fedigan LM. (2017). Trichromacy increases fruit intake rates of wild capuchins (*Cebus capucinus imitator*). *Proc Natl Acad Sci USA* 114(39): 10402–10407.

69 *As reported in an article in* Time *magazine from 1940*: If you are interested in reading the original *Time* article from 1940, see the following website: http://content.time.com/time/magazine/article/0,9171,772387,00.html.

69 *Concetta Antico is a good illustration of the genetic superiority*: Richard Roche, Sean Commins, Francesca Farina. (2018). *Why Science Needs Art: From Historical to Modern Day Perspectives*. Abingdon: Routledge.

70 *Have you ever wondered why your animal companion*: If you'd like to read more about how genetics dictates specific individual dietary requirements, then read Sharon Moalem. (2016). *The DNA Restart: Unlock Your Personal Genetic Code to Eat for Your Genes, Lose Weight, and Reverse Aging*. New York: Rodale. The following paper also provides a good overview on the genetics of vitamin C production: Drouin G, Godin JR, Pagé B. (2011). The genetics of vitamin C loss in vertebrates. *Curr Genomics* 12(5): 371–378.

71 *Every other mammal on the planet*: Nishikimi M, Kawai T, Yagi K. (1992). Guinea pigs possess a highly mutated gene for L-gulono-gamma-lactone oxidase, the key enzyme for L-ascorbic acid biosynthesis missing in this species. *J Biol Chem* 267(30): 21967–21972; Cui J, Yuan X, Wang L, Jones G, Zhang S. (2011). Recent loss of vitamin C biosynthesis ability in bats. *PLoS One* 6(11): e27114.

72 *Research on the behavior of one of our primate relatives*: Melin AD, Chiou KL, Walco ER, Bergstrom ML, Kawamura S, Fedigan LM. (2017). Trichromacy increases fruit intake rates of wild capuchins (*Cebus capucinus imitator*). *Proc Natl Acad Sci USA* 114(39): 10402–10407.

72 *Other research on captive rhesus macaques*: Melin AD, Kline DW, Hickey CM, Fedigan LM. (2013). Food search through the eyes of a monkey: A functional substitution approach for assessing the ecology of primate color vision. *Vision Res* 86: 87–96; Nevo O, Valenta K, Razafimandimby D, Melin AD, Ayasse M, Chapman CA. (2018). Frugivores and the evolution of fruit colour. *Biol Lett* 14(9); Michael Price. You can thank your fruit-hunting ancestors for your color vision. *Science*, February 19, 2017.

76 *Dr. Rita Levi-Montalcini was out of a job*: Chao MV, Calissano P. (2013). Rita Levi-Montalcini: In memoriam. *Neuron* 77(3): 385–387; Chirchiglia D, Chirchiglia P, Pugliese D, Marotta R. (2019). The legacy of Rita Levi-Montalcini: From nerve growth factor to neuroinflammation. *Neuroscientist*. doi: 10.1177/1073858419827273; Federico A. (2013). Rita Levi-Montalcini, one of the most prominent Italian personalities of the twentieth century. *Neurol Sci* 34(2): 131–133.

79 *When these tracheal mites*: Cepero A, Martín-Hernández R, Prieto L,

Gómez-Moracho T, Martínez-Salvador A, Bartolomé C, Maside X, Meana A, Higes M. (2015). Is *Acarapis woodi* a single species? A new PCR protocol to evaluate its prevalence. *Parasitol Res* 114(2): 651–658; Ochoa R, Pettis JS, Erbe E, Wergin WP. (2005). Observations on the honey bee tracheal mite *Acarapis woodi* (Acari: Tarsonemidae) using low-temperature scanning electron microscopy. *Exp Appl Acarol* 35(3): 239–249.

80 *Today we call that mysterious and previously unknown*: Manca A, Capsoni S, Di Luzio A, Vignone D, Malerba F, Paoletti F, Brandi R, Arisi I, Cattaneo A, Levi-Montalcini R. (2012). Nerve growth factor regulates axial rotation during early stages of chick embryo development. *Proc Natl Acad Sci USA* 109(6): 2009–2014; Levi-Montalcini R. (2000). From a home-made laboratory to the Nobel Prize: An interview with Rita Levi Montalcini. *Int J Dev Biol* 44(6): 563–566.

81 *Other important neurotrophins that have been identified*: Götz R, Köster R, Winkler C, Raulf F, Lottspeich F, Schartl M, Thoenen H. (1994). Neurotrophin-6 is a new member of the nerve growth factor family. *Nature* 372(6503): 266–269; Skaper SD. (2017). Nerve growth factor: A neuroimmune crosstalk mediator for all seasons. *Immunology* 151(1): 1–15.

81 *The amount of neurotrophins present within our bodies*: De Assis GG, Gasanov EV, de Sousa MBC, Kozacz A, Murawska-Cialowicz E. (2018). Brain derived neutrophic factor, a link of aerobic metabolism to neuroplasticity. *J Physiol Pharmacol* 69(3); Mackay CP, Kuys SS, Brauer SG. (2017). The effect of aerobic exercise on brain-derived neurotrophic factor in people with neurological disorders: A systematic review and meta-analysis. *Neural Plast*. doi: 10.1155/2017/4716197.

82 *"the sweetest and most mongrel dog I ever saw"*: Susan Tyler Hitchcock. (2004). *Rita Levi-Montalcini (Women in Medicine)*. Langhorne, PA: Chelsea House; Yount L. (2009). *Rita Levi-Montalcini: Discoverer of Nerve Growth Factor (Makers of Modern Science)*. Langhorne, PA: Chelsea House.

82 *More than forty years later, for the work that started*: Bradshaw RA. (2013). Rita Levi-Montalcini (1909–2012). *Nature* 493(7432): 306; Levi-Montalcini R, Knight RA, Nicotera P, Nisticó G, Bazan N, Melino G. (2011). Rita's 102!! *Mol Neurobiol* 43(2): 77–79; Chao MV, Calissano P. (2013). Rita Levi-Montalcini: In memoriam. *Neuron* 77(3): 385–387.

82 *The human brain is massive and metabolically expensive*: Lennie P. (2003). The cost of cortical computation. *Curr Biol* 13(6): 493–497; Magistretti PJ, Allaman I. (2015). A cellular perspective on brain energy metabolism and functional imaging. *Neuron* 86(4): 883–901.

83 *Evidence from various neuroscientific studies*: Rodríguez-Iglesias N, Sierra A, Valero J. (2019). Rewiring of memory circuits: Connecting adult newborn neurons with the help of microglia. *Front Cell Dev Biol* 7: 24.

83 *Some of the latest neuroscience research is implicating*: Paolicelli RC, Bolasco G, Pagani F, Maggi L, Scianni M, Panzanelli P, Giustetto M, Ferreira TA, Guiducci E, Dumas L, Ragozzino D, Gross CT. (2011). Synaptic pruning by microglia is necessary for normal brain development. *Science* 333(6048): 1456–1458; Salter MW, Stevens B. (2017). Microglia emerge as central players in brain disease. *Nat Med* 23(9): 1018–1027.

84 *This theory had been around the edges of the scientific community*: Weinhard L, di Bartolomei G, Bolasco G, Machado P, Schieber NL, Neniskyte U, Exiga M, Vadisiute A, Raggioli A, Schertel A, Schwab Y, Gross CT. (2018). Microglia remodel synapses by presynaptic trogocytosis and spine head filopodia induction. *Nat Commun* 9(1): 1228.

84 *We now think that microglia play a role*: van der Poel M, Ulas T, Mizee MR, Hsiao CC, Miedema SSM, Adelia, Schuurman KG, Helder B, Tas SW, Schultze JL, Hamann J, Huitinga I. (2019). Transcriptional profiling of human microglia reveals grey-white matter heterogeneity and multiple sclerosis-associated changes. *Nat Commun* 10(1): 1139; Zrzavy T, Hametner S, Wimmer I, Butovsky O, Weiner HL, Lassmann H. (2017). Loss of "homeostatic" microglia and patterns of their activation in active multiple sclerosis. *Brain* 140(7): 1900–1913.

85 *Now that they've piqued our interests from a disease perspective*: For more information on the emerging role that microglia may play in the development of a variety of diseases, see the following articles: Felsky D, Roostaei T, Nho K, Risacher SL, Bradshaw EM, Petyuk V, Schneider JA, Saykin A, Bennett DA, De Jager PL. (2019). Neuropathological correlates and genetic architecture of microglial activation in elderly human brain. *Nat Commun* 10(1): 409; Inta D, Lang UE, Borgwardt S, Meyer-Lindenberg A, Gass P. (2017). Microglia activation and schizophrenia: Lessons from the effects of minocycline on postnatal neurogenesis, neuronal survival and synaptic pruning. *Schizophr Bull* 43(3): 493–496; Regen F, Hellmann-Regen J, Costantini E, Reale M. (2017). Neuroinflammation and Alzheimer's disease: Implications for microglial activation. *Curr Alzheimer Res* 14(11): 1140–1148; Sellgren CM, Gracias J, Watmuff B, Biag JD, Thanos JM, Whittredge PB, Fu T, Worringer K, Brown HE, Wang J, Kaykas A, Karmacharya R, Goold CP, Sheridan SD, Perlis RH. (2019). Increased synapse elimination by

microglia in schizophrenia patient-derived models of synaptic pruning. *Nat Neurosci* 22(3): 374–385.

85 *A relatively large and recent postmortem study*: Meltzer A, Van de Water J. (2017). The role of the immune system in autism spectrum disorder. *Neuropsychopharmacology* 42(1): 284–298.

88 *Back in the 1960s, there was a lot of emphasis on the Y chromosome*: Borga-onkar DS, Murdoch JL, McKusick VA, Borkowf SP, Money JW, Rob-inson BW. (1968). The YY syndrome. *Lancet* 2(7565): 461–462; Nielsen J, Stürup G, Tsuboi T, Romano D. (1969). Prevalence of the XYY syn-drome in an institution for psychologically abnormal criminals. *Acta Psychiatr Scand* 45(4): 383–401; Fox RG. (1971). The XYY offender: A modern myth? *Journal of Crim Law and Crimonol* 62(1): 59–73.

90 *What we do know about the* MAOA *gene*: Godar SC, Fite PJ, McFarlin KM, Bortolato M. (2016). The role of monoamine oxidase A in aggres-sion: Current translational developments and future challenges. *Prog Neuropsychopharmacol Biol Psychiatry* 69: 90–100.

90 *This is exactly what a Dutch geneticist*: Brunner HG, Nelen M, Breake-field XO, Ropers HH, van Oost BA. (1993). Abnormal behavior as-sociated with a point mutation in the structural gene for monoamine oxidase A. *Science* 262(5133): 578–580.

91 *Since Brunner's original report*: Godar SC, Bortolato M, Castelli MP, Casti A, Casu A, Chen K, Ennas MG, Tambaro S, Shih JC. (2014). The aggression and behavioral abnormalities associated with monoamine oxidase A deficiency are rescued by acute inhibition of serotonin re-uptake. *J Psychiatr Res* 56: 1–9; Godar SC, Bortolato M, Frau R, Dousti M, Chen K, Shih JC. (2011). Maladaptive defensive behaviours in monoamine oxidase A-deficient mice. *Int J Neuropsychopharmacol* 14(9): 1195–1207; Scott AL, Bortolato M, Chen K, Shih JC. (2008). Novel monoamine oxidase A knock out mice with human-like spontaneous mutation. *Neuroreport* 19(7): 739–743.

91 *Mischaracterized as the "warrior gene"*: McDermott R, Tingley D, Cowden J, Frazzetto G, Johnson DD. (2009). Monoamine oxidase A gene (*MAOA*) predicts behavioral aggression following provocation. *Proc Natl Acad Sci USA* 106(7): 2118–2123; Chester DS, DeWall CN, Derefinko KJ, Estus S, Peters JR, Lynam DR, Jiang Y. (2015). Mono-amine oxidase A (*MAOA*) genotype predicts greater aggression through impulsive reactivity to negative affect. *Behav Brain Res* 283: 97–101; González-Tapia MI, Obsuth I. (2015). "Bad genes" and criminal re-sponsibility. *Int J Law Psychiatry* 39: 60–71.

92 *He also read an article that described the* MAOA *gene*: For more information about the *MAOA* gene, see Hunter P. (2010). The psycho gene. *EMBO Rep* 11(9): 667–669.

93 *I also referred him to a research paper*: Palmer EE, Leffler M, Rogers C, Shaw M, Carroll R, Earl J, Cheung NW, Champion B, Hu H, Haas SA, Kalscheuer VM, Gecz J, Field M. (2016). New insights into Brunner syndrome and potential for targeted therapy. *Clin Genet* 89(1): 120–127.

95 *Everyone is born with good and bad traits, that's what makes us human*: This quote is from Sarah Anne Murphy's graduate thesis, entitled, "Born to Rage?: A Case Study of the Warrior Gene," which can be found at the website https://wakespace.lib.wfu.edu/bitstream/handle/10339/37295 /Murphy_wfu_0248M_10224.pdf.

4. Stamina: How Women Outlast Men

98 *Our life expectancies have been increasing significantly*: Adair T, Kippen R, Naghavi M, Lopez AD. (2019). The setting of the rising sun? A recent comparative history of life expectancy trends in Japan and Australia. *PLoS One* 14(3): e0214578; GBD 2015 Mortality and Causes of Death Collaborators. (2016). Global, regional, and national life expectancy, all-cause mortality, and cause-specific mortality for 249 causes of death, 1980–2015: A systematic analysis for the Global Burden of Disease Study. *Lancet* 388(10053): 1459–1544. For an article particularly focusing on countries with long-lived individuals, including Japan, see Marina Pitofsky. What countries have the longest life expectancies. *USA Today*, July 27, 2018. https://eu.usatoday.com/story/news/2018/07 /27/life-expectancies-2018-japan-switzerland-spain/848675002/.

98 *Even Afghanistan, a country with one of the lowest life-expectancy levels*: For more data and information, see the 2010 Afghanistan Mortality Survey at the United States Agency for International Development website: https:// www.usaid.gov/sites/default/files/documents/1871/Afghanistan%20 Mortality%20Survey%20Key%20Findings.pdf.

98 *That's still significantly longer than what a seventeenth-century Londoner*: For more on this topic, see Griffin JP. (2008). Changing life expectancy throughout history. *J R Soc Med* 101(12): 577.

99 *In a book published in 1662*: Benjamin B, Clarke RD, Beard RE, Brass W. (1963). A discussion on demography. *Proc R Soc Lon Series B Bio* 159(74): 38–65.

99 *During the Great Plague of London*: John Kelly. (2005). *The Great Mortality: An Intimate History of the Black Death, the Most Devastating Plague*

of All Time. New York: Harper; Greenberg SJ. (1997). The "dreadful visitation": Public health and public awareness in seventeenth-century London. *Bull Med Libr Assoc* 85(4): 391–401; Raoult D, Mouffok N, Bitam I, Piarroux R, Drancourt M. (2013). Plague: History and contemporary analysis. *J Infect* 66(1): 18–26. For more about bubonic plague mortality in London between the fourteenth and seventeenth centuries, see Twigg G. (1992). Plague in London: Spatial and temporal aspects of mortality. https://www.history.ac.uk/cmh/epitwig .html.

99 *To help predict and track untimely deaths:* For more on the important role that searchers played in the history of London, see Munkhoff R. (1999). Searchers of the dead: Authority, marginality, and the interpretation of plague in England, 1574–1665. *Gend Hist* 11(1): 1–29.

99 *The documents, known as the Bills of Mortality:* Morabia A. (2013). Epidemiology's 350th anniversary: 1662–2012. *Epidemiology* 24(2): 179–183; Slauter W. (2011). Write up your dead. *Med Hist* 17(1): 1–15.

100 *The seventeenth-century Englishman Edmond Halley:* Bellhouse DR. (2011). A new look at Halley's life table. *J Royal Stat Soc Series A* 174(3): 823–832; Halley E. (1693): An estimate of the degrees of the mortality of mankind, drawn from curious tables of the births and funerals at the city of Breslaw; with an attempt to ascertain the price of annuities upon lives. *Phil Trans Roy Soc London* 17: 596–610; Mary Virginia Fox. (2007). *Scheduling the Heavens: The Story of Edmond Halley.* Greensboro, NC: Morgan Reynolds; John Gribbin, Mary Gribbin. (2017). *Out of the Shadow of a Giant: Hooke, Halley, and the Birth of Science.* New Haven, CT: Yale University Press.

101 *Halley's paper was a highly significant contribution:* Anders Hald. (2003). *A History of Probability and Statistics and Their Applications Before 1750.* Hoboken, NJ: John Wiley and Sons.

101 *The significance of Graunt's findings regarding the survival of women:* Peter Sprent, Nigel C. Smeeton. (2007). *Applied Nonparametric Statistical Methods.* Boca Raton, FL: CRC Press.

101 *No matter where you look around the world:* There's a lot of published literature on female longevity. If you'd like to read more about this topic, these few papers are a good place to begin: Marais GAB, Gaillard JM, Vieira C, Plotton I, Sanlaville D, Gueyffier F, Lemaitre JF. (2018). Sex gap in aging and longevity: Can sex chromosomes play a role? *Biol Sex Differ* 9(1): 33; Pipoly I, Bokony V, Kirkpatrick M, Donald PF, Szekely T, Liker A. (2015). The genetic sex-determination system predicts adult sex ratios

in tetrapods. *Nature* 527(7576): 91–94; Austad SN, Fischer KE. (2016). Sex differences in lifespan. *Cell Metab* (6): 1022–1033.

102 *The story of Marguerite de La Rocque*: There is ongoing debate regarding the exact island on which Marguerite was abandoned. The most likely location is Belle Isle, though some have suggested alternative locations. If you would like to read a historical novel that is roughly based on Marguerite's story, see the following book: Elizabeth Boyer. (1975). *Marguerite de La Roque: A Story of Survival*. Novelty, OH: Veritie Press.

104 *The story of the Donner Party*: Much has been written about the Donner Party's fateful journey. For a few accounts, see the following: Donald L. Hardesty. (2005). *The Archaeology of the Donner Party*. Reno: University of Nevada Press; Grayson DK. (1993). Differential mortality and the Donner Party disaster. *Evol Anthropol* 2: 151–159.

105 *Life expectancy for Ukrainians prior to the calamity*: Estimated life expectancies were published here: France Meslé, Jacques Vallin (2012). *Mortality and Causes of Death in 20th-Century Ukraine*. New York: Springer Science and Business Media. The work of Meslé and Vallin is considered to be based on solid historical and statistical references that are the most reliable for that era. For more information, see Zarulli V, Barthold Jones JA, Oksuzyan A, Lindahl-Jacobsen R, Christensen K, Vaupel JW. (2018). Women live longer than men even during severe famines and epidemics. *Proc Natl Acad Sci USA* 115(4): E832–E840.

105 *Genetic males might have more muscle mass*: For more information on some of the physical differences between genetic females and males, see Ellen Casey, Joel Press J, Monica Rho M. (2016). *Sex Differences in Sports Medicine*. New York: Springer Publishing.

106 *Data from the nineteenth and early twentieth centuries*: Lindahl-Jacobsen R, Hanson HA, Oksuzyan A, Mineau GP, Christensen K, Smith KR. (2013). The male-female health-survival paradox and sex differences in cohort life expectancy in Utah, Denmark, and Sweden 1850–1910. *Ann Epidemiol* 23(4): 161–166.

106 *Men do outnumber women in the occupations that are the most dangerous*: For a breakdown of the causes of fatal occupational injuries in the United States, see Clougherty JE, Souza K, Cullen MR. (2010). Work and its role in shaping the social gradient in health. *Ann N Y Acad Sci* 1186: 102–124; Bureau of Labor Statistics. Census of fatal occupational injuries summary, 2017. https://www.bls.gov/news.release/cfoi.nr0.htm.

106 *Yet a study from Germany also found*: Austad SN, Fischer KE. (2016). Sex differences in lifespan. *Cell Metab* 23(6): 1022–1033; Luy M. (2003).

Causes of male excess mortality: Insights from cloistered populations. *Pop and Dev Review* 29(4): 647–676.

108 *We are now discovering that of the one thousand genes*: Tukiainen T, Villani AC, Yen A, Rivas MA, Marshall JL, Satija R, Aguirre M, Gauthier L, Fleharty M, Kirby A, Cummings BB, Castel SE, Karczewski KJ, Aguet F, Byrnes A; GTEx Consortium; Laboratory, Data Analysis and Coordinating Center (LDACC)—Analysis Working Group; Statistical Methods groups—Analysis Working Group; Enhancing GTEx (eGTEx) groups; NIH Common Fund; NIH/NCI; NIH/NHGRI; NIH/NIMH; NIH/NIDA; Biospecimen Collection Source Site—NDRI; Biospecimen Collection Source Site—RPCI; Biospecimen Core Resource—VARI; Brain Bank Repository—University of Miami Brain Endowment Bank; Leidos Biomedical—Project Management; ELSI Study; Genome Browser Data Integration &Visualization—EBI; Genome Browser Data Integration and Visualization—UCSC Genomics Institute, University of California Santa Cruz; Lappalainen T, Regev A, Ardlie KG, Hacohen N, MacArthur DG. (2017). Landscape of X chromosome inactivation across human tissues. *Nature* 550(7675): 244–248; Snell DM, Turner JMA. (2018). Sex chromosome effects on male-female differences in mammals. *Curr Biol* 28(22): R1313–R1324; Raznahan A, Parikshak NN, Chandran V, Blumenthal JD, Clasen LS, Alexander-Bloch AF, Zinn AR, Wangsa D, Wise J, Murphy DGM, Bolton PF, Ried T, Ross J, Giedd JN, Geschwind DH. (2018). Sex-chromosome dosage effects on gene expression in humans. *Proc Natl Acad Sci USA* 115(28): 7398–7403; Balaton BP, Dixon-McDougall T, Peeters SB, Brown CJ. (2018). The eXceptional nature of the X chromosome. *Hum Mol Genet* 27(R2): R242–R249.

110 *More than ten thousand years ago, our circumstances*: Marcel Mazoyer, Laurence Roudart. (2006). *A History of World Agriculture: From the Neolithic Age to the Current Crisis*. New York: Monthly Review Press.

110 *Cooking with fire, which we mastered long before*: Attwell L, Kovarovic K, Kendal J. (2015). Fire in the Plio-Pleistocene: The functions of hominin fire use, and the mechanistic, developmental and evolutionary consequences. *J Anthropol Sci* 93: 1–20; Gowlett JA. (2016). The discovery of fire by humans: A long and convoluted process. *Philos Trans R Soc Lond B Biol Sci* 371: 1696.

115 *In summer 1771, that's exactly what started happening*: Dribe M, Olsson M, Svensson P. (2015). Famines in the Nordic countries, AD 536–1875. *Lund Papers in Economic History* 138: 1–41; Zarulli V, Barthold Jones JA, Oksuzyan A, Lindahl-Jacobsen R, Christensen K, Vaupel JW.

(2018). Women live longer than men even during severe famines and epidemics. *Proc Natl Acad Sci USA* 115(4): E832–E840.

116 *It included death registration and census data*: The data is from Zarulli V. Biology makes women and girls survivors. July 15, 2018. http://sciencenordic .com/biology-makes-women-and-girls-survivors; as well as the paper that Dr. Virginia Zarulli coauthored and published in the *PNAS*: Zarulli V, Barthold Jones JA, Oksuzyan A, Lindahl-Jacobsen R, Christensen K, Vaupel JW. (2018). Women live longer than men even during severe famines and epidemics. *Proc Natl Acad Sci USA* 115(4): E832–E840.

120 *Domesticated potatoes* (Solanum tuberosum L.) *are just as essential*: Andrew F. Smith. (2011). *Potato: A Global History*. London: Reaktion Books; Machida-Hirano R. (2015). Diversity of potato genetic resources. *Breed Sci* 65(1): 26–40; Camire ME, Kubow S, Donnelly DJ. (2009). Potatoes and human health. *Crit Rev Food Sci Nutr* 49(10): 823–840.

122 *Domesticated potato plants, on the other hand*: Comai L. (2005). The advantages and disadvantages of being polyploid. *Nat Rev Genet* (11): 836–846; Tanvir-Ul-Hassan Dar, Reiaz-Ul Rehman. (2017). *Polyploidy: Recent Trends and Future Perspectives*. New York: Springer.

128 *Children with Hunter syndrome can end up*: Muenzer J, Jones SA, Tylki-Szymańska A, Harmatz P, Mendelsohn NJ, Guffon N, Giugliani R, Burton BK, Scarpa M, Beck M, Jangelind Y, Hernberg-Stahl E, Larsen MP, Pulles T, Whiteman DAH. (2017). Ten years of the Hunter Outcome Survey (HOS): Insights, achievements, and lessons learned from a global patient registry. *Orphanet J Rare Dis* 12(1): 82.

129 *When I first met him, Simon was already*: Whiteman DA, Kimura A. (2017). Development of idursulfase therapy for mucopolysaccharidosis type II (Hunter syndrome): The past, the present and the future. *Drug Des Devel Ther* 11: 2467–2480; Muenzer J, Giugliani R, Scarpa M, Tylki-Szymańska A, Jego V, Beck M. (2017). Clinical outcomes in idursulfase-treated patients with mucopolysaccharidosis type II: 3-year data from the Hunter Outcome Survey (HOS). *Orphanet J Rare Dis* 12(1): 161; Sohn YB, Cho SY, Park SW, Kim SJ, Ko AR, Kwon EK, Han SJ, Jin DK. (2013). Phase I/II clinical trial of enzyme replacement therapy with idursulfase beta in patients with mucopolysaccharidosis II (Hunter syndrome). *Orphanet J Rare Dis* 8: 42.

135 *She won the Moab 240, a 238.3-mile footrace*: If you would like to read more about Courtney, see the following: Ariella Gintzler. How Courtney Dauwalter won the Moab 240 outright: The 32-year-old gapped second place (and first male) by 10 hours. *Trail Runner Magazine*, October

18, 2017. https://trailrunnermag.com/people/news/courtney-dauwalter
-wins-moab-240.html.

135 *"Sometimes when I leave my house I don't even know"*: This quote is from the
following article: Taylor Rojek. There's no stopping ultrarunner Courtney Dauwalter: The 33-year-old ultrarunner is smashing records—and
she doesn't plan on slowing down. *Runner's World*, August 3, 2018.

136 *The Montane Spine Race is another brutal ultra-endurance competition*:
Angie Brown. The longer the race, the stronger we get. *BBC Scotland*,
January 17, 2019. https://www.bbc.com/news/uk-scotland-edinburgh
-east-fife-46906365.

136 *"All these guys will go out hot, and hours later I catch them"*: This quote is
from the following article: Meaghan Brown. The longer the race, the
stronger we get. *Outside*, April 11, 2017. https://www.outsideonline
.com/2169856/longer-race-stronger-we-get.

137 *"When I was coming into the race, I thought that maybe I could go for the
women's podium"*: This quote is from the following article, where you can
also read more about Kolbinger's win: Helen Pidd. Cancer researcher
becomes first women to win 4,000km cycling race. *The Guardian*. August 6, 2019. https://www.theguardian.com/sport/2019/aug/06/fiona
-kolbinger-first-woman-win-transcontinental-cycling-race.

5. Superimmunity: The Costs and Benefits of Genetic Superiority

139 *Smallpox has easily been the source*: Ghio AJ. (2017). Particle exposure and
the historical loss of Native American lives to infections. *Am J Respir
Crit Care Med* 195(12): 1673; Shchelkunov SN. (2011). Emergence and
reemergence of smallpox: The need for development of a new generation
smallpox vaccine. *Vaccine* 29(Suppl 4): D49–53; Voigt EA, Kennedy
RB, Poland GA. (2016). Defending against smallpox: A focus on vaccines. *Expert Rev Vaccines* 15(9): 1197–1211.

139 *The Intensified Smallpox Eradication Campaign*: For more on this topic,
see Frank Fenner, Donald A. Henderon, Isao Arita, Zdeněk Ježek, Ivan
D. Ladnyi. (1988). *Smallpox and Its Eradication*. Geneva: World Health
Organization; Jack W. Hopkins. (1989). *The Eradication of Smallpox:
Organizational Learning and Innovation in International Health*. Boul-
der, CO: Westview Press.

140 *Then the flu-like symptoms begin*: Reardon S. (2014). Infectious diseases:
Smallpox watch. *Nature* 509(7498): 22–24.

141 *In 1980 the WHO officially declared smallpox eradicated*: For more on the
WHO's astonishing accomplishment of completely eradicating small-

pox worldwide, see World Health Organization. The Smallpox Eradication Programme—SEP (1966–1980). https://www.who.int/features/2010/smallpox/en/.

142 *The grand narrative of the outstanding scientific*: D'Amelio E, Salemi S, D'Amelio R. (2016). Anti-infectious human vaccination in historical perspective. *Int Rev Immunol* 35(3): 260–290; Hajj Hussein I, Chams N, Chams S, El Sayegh S, Badran R, Raad M, Gerges-Geagea A, Leone A, Jurjus A. (2015). Vaccines through centuries: Major cornerstones of global health. *Front Public Health* 3: 269.

143 *For his work on cuckoo birds, Jenner was elected*: Riedel S. (2005). Edward Jenner and the history of smallpox and vaccination. *Proc (Bayl Univ Med Cent)* 18(1): 21–25.

143 *There are a few stories about how Jenner*: Damaso CR. (2018). Revisiting Jenner's mysteries, the role of the Beaugency lymph in the evolutionary path of ancient smallpox vaccines. *Lancet Infect Dis* 18(2): e55–e63.

143 *Both cowpox and smallpox are caused*: Mucker EM, Hartmann C, Hering D, Giles W, Miller D, Fisher R, Huggins J. (2017). Validation of a *pan*-orthopox real-time PCR assay for the detection and quantification of viral genomes from nonhuman primate blood. *Virol J* 14(1): 210.

144 *So, Jenner used James Phipps*: See the following paper for more information on the debate regarding the ethical use of human research subjects: Davies H. (2007). Ethical reflections on Edward Jenner's experimental treatment. *J Med Ethics* 33(3): 174–176.

144 *The father of immunology had his work rejected*: Riedel S. (2005). Edward Jenner and the history of smallpox and vaccination. *Proc (Bayl Univ Med Cent)* 18(1): 21–25.

145 *He finally published his work, titled*: Jenson AB, Ghim SJ, Sundberg JP. (2016). An inquiry into the causes and effects of the variolae (or Cowpox. 1798). *Exp Dermatol* 25(3): 178–180.

145 *Ultimately, he was even awarded grants*: For more information on Jenner, see the London School of Hygiene and Tropical Medicine. Edward Jenner (1749–1823). https://www.lshtm.ac.uk/aboutus/introducing/history/frieze/edward-jenner.

145 *It took just ten short years after Jenner's initial experiments*: Rusnock A. (2009). Catching cowpox: The early spread of smallpox vaccination, 1798–1810. *Bull Hist Med* 83(1): 17–36.

146 *Lady Mary Montagu was born on May 26, 1689*: Dinc G, Ulman YI. (2007). The introduction of variolation "A La Turca" to the West by Lady Mary Montagu and Turkey's contribution to this. *Vaccine* 25(21):

4261–4265; Rathbone J. (1996). Lady Mary Wortley Montague's contribution to the eradication of smallpox. *Lancet* 347(9014): 1566.

147 *"A propos of distempers, I am going to tell you a thing"*: Barnes D. (2012). The public life of a woman of wit and quality: Lady Mary Wortley Montagu and the vogue for smallpox inoculation. *Fem Stud* 38(2): 330–362; Weiss RA, Esparza J. (2015). The prevention and eradication of smallpox: A commentary on Sloane (1755) "An account of inoculation." *Philos Trans R Soc Lond B Biol Sci* 370 (1666).

148 *What Lady Montagu probably didn't know*: Dinc G, Ulman YI. (2007). The introduction of variolation "A La Turca" to the West by Lady Mary Montagu and Turkey's contribution to this. *Vaccine* 25(21): 4261–4265; Simmons BJ, Falto-Aizpurua LA, Griffith RD, Nouri K. (2015). Smallpox: 12,000 years from plagues to eradication: A dermatologic ailment shaping the face of society. *JAMA Dermatol* 151(5): 521.

148 *Females, when immunologically provoked*: Flanagan KL, Fink AL, Plebanski M, Klein SL. (2017). Sex and gender differences in the outcomes of vaccination over the life course. *Annu Rev Cell Dev Biol* 33: 577–599; Klein SL, Pekosz A. (2014). Sex-based biology and the rational design of influenza vaccination strategies. *J Infect Dis* 3: S114–119.

149 *"I am patriot enough to take the pains"*: Weiss RA, Esparza J. (2015). The prevention and eradication of smallpox: A commentary on Sloane (1755) "An account of inoculation." *Philos Trans R Soc Lond B Biol Sci* 370: 1666.

150 *Maitland was given a royal license to perform an experimental trial*: Stone AF, Stone WD. (2002). Lady Mary Wortley Montagu: Medical and religious controversy following her introduction of smallpox inoculation. *J Med Biogr* 10(4): 232–236.

151 *A few articles were published in the* Philosophical Transactions: Weiss RA, Esparza J. (2015). The prevention and eradication of smallpox: A commentary on Sloane (1755) "An account of inoculation." *Philos Trans R Soc Lond B Biol Sci* 370: 1666.

152 *Using the technique first developed by his father*: Although never formally trained as either a scientist or a physician, Sutton variolated thousands of people against smallpox and made many interesting observations. For details of his intriguing life story, see Boylston A. (2012). Daniel Sutton, a forgotten 18th century clinician scientist. *J R Soc Med* 105(2): 85–87.

152 *"The air of the palace was infected"*: This quote is from Weiss RA, Esparza J. (2015). The prevention and eradication of smallpox: A commentary on Sloane (1755) "An account of inoculation." *Philos Trans R Soc Lond B Biol Sci* 370: 1666.

153 *B cells carry about one hundred thousand identical copies*: Bruce Alberts, Alexander Johnson, Julian Lewis, Martin Raff, Keith Roberts, and Peter Walter. (2002). *Molecular Biology of the Cell.* 4th ed. New York: Garland Science; Li J, Yin W, Jing Y, Kang D, Yang L, Cheng J, Yu Z, Peng Z, Li X, Wen Y, Sun X, Ren B, Liu C. (2019). The coordination between B cell receptor signaling and the actin cytoskeleton during B cell activation. *Front Immunol* 9: 3096.

155 *Women usually experience more pain and side effects*: Klein SL, Pekosz A. (2014). Sex-based biology and the rational design of influenza vaccination strategies. *J Infect Dis* 209 Suppl 3: S114–9; Klein SL, Marriott I, Fish EN. (2015). Sex-based differences in immune function and responses to vaccination. *Trans R Soc Trop Med Hyg* 109(1): 9–15.

156 *Geographically speaking, the cities of Atlanta, Georgia*: Nalca A, Zumbrun EE. (2010). ACAM2000: The new smallpox vaccine for United States Strategic National Stockpile. *Drug Des Devel Ther* 4: 71–79; Nagata LP, Irwin CR, Hu WG, Evans DH. (2018). Vaccinia-based vaccines to biothreat and emerging viruses. *Biotechnol Genet Eng Rev* 34(1): 107–121; Petersen BW, Damon IK, Pertowski CA, Meaney-Delman D, Guarnizo JT, Beigi RH, Edwards KM, Fisher MC, Frey SE, Lynfield R, Willoughby RE. (2015). Clinical guidance for smallpox vaccine use in a postevent vaccination program. *MMWR Recomm Rep* 64(RR-02): 1–26; Habeck M. (2002). UK awards contract for smallpox vaccine. *Lancet Infect Dis* 2(6): 321; Stamm LV. (2015). Smallpox redux? *JAMA Dermatol* 151(1): 13–14; Wiser I, Balicer RD, Cohen D. (2007). An update on smallpox vaccine candidates and their role in bioterrorism related vaccination strategies. *Vaccine* 25(6): 976–984.

156 *Some of my own research has involved developing*: For more on the history of the black death, in which I delve in greater detail, see one of my previous books: Sharon Moalem with Jonathan M. Prince. (2007). *Survival of the Sickest: A Medical Maverick Discovers Why We Need Disease.* New York: William Morrow.

157 *The killing potential of Y. pestis is amplified*: Many bacterial and fungal pathogens that infect humans and animals and cause disease rely on the availability and acquisition of the metal iron. Interestingly, many of these or related microbes will also infect plants, insects, and other vertebrate species as well. For more information and a list of other bacterial organisms that depend on iron for their pathogenicity, see the following articles: Moalem S, Weinberg ED, Percy ME. (2004). Hemochromatosis and the enigma of misplaced iron: Implications for infectious disease and survival. *Biometals* 17(2): 135–139; Holden VI,

Bachman MA. (2015). Diverging roles of bacterial siderophores during infection. *Metallomics* 7(6): 986–995; Lyles KV, Eichenbaum Z. (2018). From host heme to iron: The expanding spectrum of heme degrading enzymes used by pathogenic bacteria. *Front Cell Infect Microbiol* 8: 198; Nevitt T. (2011). War-Fe-re: Iron at the core of fungal virulence and host immunity. *Biometals* 24(3): 547–558; Rakin A, Schneider L, Podladchikova O. (2012). Hunger for iron: The alternative siderophore iron scavenging systems in highly virulent *Yersinia*. *Front Cell Infect Microbiol* 2: 151.

157 *Giving* Y. pestis *the genetic ability to acquire*: Perry RD, Fetherston JD. (2011). Yersiniabactin iron uptake: Mechanisms and role in *Yersinia pestis* pathogenesis. *Microbes Infect* 13(10): 808–817.

159 *Genetic males born with X-linked agammaglobulinemia*: Berglöf A, Turunen JJ, Gissberg O, Bestas B, Blomberg KE, Smith CI. (2013). Agammaglobulinemia: Causative mutations and their implications for novel therapies. *Expert Rev Clin Immunol* 9(12): 1205–1221.

160 *Some of the PRRs—such as the toll-like receptor genes*: Souyris M, Cenac C, Azar P, Daviaud D, Canivet A, Grunenwald S, Pienkowski C, Chaumeil J, Mejía JE, Guéry JC. (2018). TLR7 escapes X chromosome inactivation in immune cells. *Sci Immunol* 3(19).

162 *Sadly, we've seen what happens when they are not functioning in cases like David Vetter*: For more information about David Vetter's life, see the following papers: Berg LJ. (2008). The "bubble boy" paradox: An answer that led to a question. *J Immunol* 181(9): 5815–5816; Hollander SA, Hollander EJ. (2018). The boy in the bubble and the baby with the Berlin heart: The dangers of "bridge to decision" in pediatric mechanical circulatory support. *ASAIO J* 64(6): 831–832. There's also a list of articles and summaries about his story and condition at the Immune Deficiency Foundation website: https://primaryimmune.org/living-pi-explaining -pi-others/story-david.

163 *This distinction may necessitate men receiving*: Klein SL, Pekosz A. (2014). Sex-based biology and the rational design of influenza vaccination strategies. *J Infect Dis* 3: S114–119.

163 *Around five million people worldwide have the disorder*: Rider V, Abdou NI, Kimler BF, Lu N, Brown S, Fridley BL. (2018). Gender bias in human systemic lupus erythematosus: A problem of steroid receptor action? *Front Immunol* 9: 611.

164 *"The wolf, I'm afraid, is inside tearing up the place"*: Donald E. Thomas. (2014). *The Lupus Encyclopedia: A Comprehensive Guide for Patients and Families*. Baltimore: Johns Hopkins University Press.

165 *The National Institutes of Health (NIH) estimates*: For more information on autoimmune diseases, see the following NIH website: https://www.niehs.nih.gov/health/materials/autoimmune_diseases_508.pdf.

165 *For the most part, autoimmune conditions predominantly affect females*: Chiaroni-Clarke RC, Munro JE, Ellis JA. (2016). Sex bias in paediatric autoimmune disease—not just about sex hormones? *J Autoimmun* 69: 12–23; Gary S. Firestein, Ralph C. Budd, Sherine E. Gabriel, Iain B. McInnes, James R. O'Dell. (2017). *Kelley and Firestein's Textbook of Rheumatology*. New York: Elsevier.

165 *In 1900, Paul Ehrlich, who would be awarded the Nobel Prize*: Mackay IR. (2010). Travels and travails of autoimmunity: A historical journey from discovery to rediscovery. *Autoimmun Rev* 9(5): A251–258; Silverstein AM. (2001). Autoimmunity versus horror autotoxicus: The struggle for recognition. *Nat Immunol* 2(4): 279–281.

167 *While our B cells make antibodies to battle pathogens*: Cruz-Adalia A, Ramirez-Santiago G, Calabia-Linares C, Torres-Torresano M, Feo L, Galán-Díez M, Fernández-Ruiz E, Pereiro E, Guttmann P, Chiappi M, Schneider G, Carrascosa JL, Chichón FJ, Martínez Del Hoyo G, Sánchez-Madrid F, Veiga E. (2014). T cells kill bacteria captured by transinfection from dendritic cells and confer protection in mice. *Cell Host Microbe* 15(5): 611–622; Cruz-Adalia A, Veiga E. (2016). Close encounters of lymphoid cells and bacteria. *Front Immunol* 7: 405.

168 *Most T cells don't make it out alive*: Daley SR, Teh C, Hu DY, Strasser A, Gray DHD. (2017). Cell death and thymic tolerance. *Immunol Rev* 277(1): 9–20; Kurd N, Robey EA. (2016). T-cell selection in the thymus: A spatial and temporal perspective. *Immunol Rev* 271(1): 114–26; Xu X, Zhang S, Li P, Lu J, Xuan Q, Ge Q. (2013). Maturation and emigration of single-positive thymocytes. *Clin Dev Immunol*. doi: 10.1155/2013/282870.

168 *It all has to do with a gene called the* autoimmune regulator: Berrih-Aknin S, Panse RL, Dragin N. (2018). AIRE: A missing link to explain female susceptibility to autoimmune diseases. *Ann N Y Acad Sci* 1412(1): 21–32; Dragin N, Bismuth J, Cizeron-Clairac G, Biferi MG, Berthault C, Serraf A, Nottin R, Klatzmann D, Cumano A, Barkats M, Le Panse R, Berrih-Aknin S. (2016). Estrogen-mediated downregulation of AIRE influences sexual dimorphism in autoimmune diseases. *J Clin Invest* 126(4): 1525–1537; Passos GA, Speck-Hernandez CA, Assis AF, Mendes-da-Cruz DA. (2018). Update on Aire and thymic negative selection. *Immunology* 153(1): 10–20; Zhu ML, Bakhru P, Conley B, Nelson JS, Free M, Martin A, Starmer J, Wilson EM, Su MA. (2016).

Sex bias in CNS autoimmune disease mediated by androgen control of autoimmune regulator. *Nat Commun* 7: 11350.

172 *What we don't know yet is if the X inactivation*: Ishido N, Inoue N, Watanabe M, Hidaka Y, Iwatani Y. (2015). The relationship between skewed X chromosome inactivation and the prognosis of Graves' and Hashimoto's diseases. *Thyroid* 25(2): 256–261; Kanaan SB, Onat OE, Balandraud N, Martin GV, Nelson JL, Azzouz DF, Auger I, Arnoux F, Martin M, Roudier J, Ozcelik T, Lambert NC. (2016). Evaluation of X chromosome inactivation with respect to HLA genetic susceptibility in rheumatoid arthritis and systemic sclerosis. *PLoS One* 11(6): e0158550; Simmonds MJ, Kavvoura FK, Brand OJ, Newby PR, Jackson LE, Hargreaves CE, Franklyn JA, Gough SC. (2014). Skewed X chromosome inactivation and female preponderance in autoimmune thyroid disease: An association study and meta-analysis. *J Clin Endocrinol Metab* 99(1): E127–131.

173 *According to data compiled by the American Cancer Society*: Siegel RL, Miller KD, Jemal A. (2017). Cancer statistics, 2017. *CA Cancer J Clin* 67(1): 7–30.

173 *The latest numbers of new cases of cancers*: For a complete list of cancer types and their incidence, see the National Cancer Institute's Surveillance, Epidemiology, and End Results (SEER) Program website: https://seer.cancer.gov.

174 *Both the African and the Asian elephant have multiple copies*: Abegglen LM, Caulin AF, Chan A, Lee K, Robinson R, Campbell MS, Kiso WK, Schmitt DL, Waddell PJ, Bhaskara S, Jensen ST, Maley CC, Schiffman JD. (2015). Potential mechanisms for cancer resistance in elephants and comparative cellular response to DNA damage in humans. *JAMA* 14(17): 1850–1860.

175 *Elephants also have extra copies of a gene*: Vazquez JM, Sulak M, Chigurupati S, Lynch VJ. (2018). A zombie LIF gene in elephants is upregulated by TP53 to induce apoptosis in response to DNA damage. *Cell Rep* 24(7): 1765–1776.

176 *Women have escape from X-inactivation tumor-suppressor genes*: Dunford A, Weinstock DM, Savova V, Schumacher SE, Cleary JP, Yoda A, Sullivan TJ, Hess JM, Gimelbrant AA, Beroukhim R, Lawrence MS, Getz G, Lane AA. (2017). Tumor-suppressor genes that escape from X-inactivation contribute to cancer sex bias. *Nat Genet* 49(1): 10–16; Wainer Katsir K, Linial M. (2019). Human genes escaping X-inactivation revealed by single cell expression data. *BMC Genomics* 20(1): 201; Carrel L, Brown CJ. (2017). When the Lyon(ized chromo-

some) roars: Ongoing expression from an inactive X chromosome. *Philos Trans R Soc Lond B Biol Sci* 372(1733).

6. Well-Being: Why Women's Health Is Not Men's Health

178 *The practice of medicine was built using research*: See the following articles for an introduction to this topic: Lee H, Pak YK, Yeo EJ, Kim YS, Paik HY, Lee SK. (2018). It is time to integrate sex as a variable in preclinical and clinical studies. *Exp Mol Med* 50(7): 82; Ramirez FD, Motazedian P, Jung RG, Di Santo P, MacDonald Z, Simard T, Clancy AA, Russo JJ, Welch V, Wells GA, Hibbert B. (2017). Sex bias is increasingly prevalent in preclinical cardiovascular research: Implications for translational medicine and health equity for women; A systematic assessment of leading cardiovascular journals over a 10-year period. *Circulation* 135(6): 625–626; Rich-Edwards JW, Kaiser UB, Chen GL, Manson JE, Goldstein JM. (2018). Sex and gender differences research design for basic, clinical, and population studies: Essentials for investigators. *Endocr Rev* 39(4): 424–439; Shansky RM, Woolley CS. (2016). Considering sex as a biological variable will be valuable for neuroscience research. *J Neurosci* 36(47): 11817–11822; Weinberger AH, McKee SA, Mazure CM. (2010). Inclusion of women and gender-specific analyses in randomized clinical trials of treatments for depression. *J Womens Health* 19(9): 1727–1732.

180 *When it comes to metals such as zinc and iron*: For more information on the metal iron and the differing requirements between the genetic sexes, see National Institutes of Health. Health information: Iron. https://ods.od .nih.gov/factsheets/Iron-HealthProfessional. For zinc, the corresponding website can be found at https://ods.od.nih.gov/factsheets/Zinc -HealthProfessional/.

180 *The FDA did publish a document*: The following guidance for industry was published in 1987. For more information, see U.S. Food and Drug Association. (February 1987). Format and content of the nonclinical pharmacology/toxicology section of an application. https://www.fda .gov/downloads/Drugs/GuidanceComplianceRegulatoryInformation /Guidances/UCM079234.pdf.

180 *It went on record stating the following*: The quote is from the FDA website and can be found at https://www.fda.gov/scienceresearch/specialtopics /womenshealthresearch/ucm472932.htm.

182 *Research in the 1980s and 1990s that was looking at new drug applications*: For a historical overview and a more contemporary update regarding

the inclusion and exclusion of women from clinical trials, see Thibaut F. (2017). Gender does matter in clinical research. *Eur Arch Psychiatry Clin Neurosci* 267(4): 283–284; Zakiniaeiz Y, Cosgrove KP, Potenza MN, Mazure CM. (2016). Balance of the sexes: Addressing sex differences in preclinical research. *Yale J Biol Med* 89(2): 255–259; FDA. Gender studies in product development: Historical overview. https://www.fda.gov/ScienceResearch/SpecialTopics/WomensHealthResearch/ucm134466.htm.

182 *This discrepancy prompted the National Institutes of Health in 1993*: For more information, see National Institutes of Health. NIH policy and guidelines on the inclusion of women and minorities as subjects in clinical research. https://grants.nih.gov/grants/funding/women_min/guidelines.htm.

182 *The latest study to tackle the issue of the inclusion of women*: There's still some debate about whether women are still underrepresented in all phases of clinical trials. Much work remains to be done, especially in Phase I and Phase II, to achieve sex equality in all clinical trials for drugs and treatments that will be prescribed to both sexes. See the following study and commentary for more detailed information about this topic: Gispen-de Wied C, de Boer A. (2018). Commentary on "Gender differences in clinical registration trials; is there a real problem?" by Labots et al. *Br J Clin Pharmacol* 84(8): 1639–1640; Labots G, Jones A, de Visser SJ, Rissmann R, Burggraaf J. (2018). Gender differences in clinical registration trials: Is there a real problem? *Br J Clin Pharmacol* 84(4): 700–707; Scott PE, Unger EF, Jenkins MR, Southworth MR, McDowell TY, Geller RJ, Elahi M, Temple RJ, Woodcock J. (2018). Participation of women in clinical trials supporting FDA approval of cardiovascular drugs. *J Am Coll Cardiol* 71(18): 1960–1969.

183 *The FDA finally recognized in April 2013*: Norman JL, Fixen DR, Saseen JJ, Saba LM, Linnebur SA. (2017). Zolpidem prescribing practices before and after Food and Drug Administration required product labeling changes. *SAGE Open Med.* doi: 10.1177/2050312117707687; Booth JN III, Behring M, Cantor RS, Colantonio LD, Davidson S, Donnelly JP, Johnson E, Jordan K, Singleton C, Xie F, McGwin G Jr. (2016). Zolpidem use and motor vehicle collisions in older drivers. *Sleep Med* 20: 98–102.

183 *Even over-the-counter medications like Tylenol (acetaminophen)*: Rubin JB, Hameed B, Gottfried M, Lee WM, Sarkar M; Acute Liver Failure Study Group. (2018). Acetaminophen-induced acute liver failure is

more common and more severe in women. *Clin Gastroenterol Hepatol* 6(6): 936–946.

184 *Clinical drug trials don't always take into account*: Clayton JA, Collins FS. (2014). Policy: NIH to balance sex in cell and animal studies. *Nature* 509(7500): 282–283; Miller LR, Marks C, Becker JB, Hurn PD, Chen WJ, Woodruff T, McCarthy MM, Sohrabji F, Schiebinger L, Wetherington CL, Makris S, Arnold AP, Einstein G, Miller VM, Sandberg K, Maier S, Cornelison TL, Clayton JA. (2017). Considering sex as a biological variable in preclinical research. *FASEB J* 31(1): 29–34; Ventura-Clapier R, Dworatzek E, Seeland U, Kararigas G, Arnal JF, Brunelleschi S, Carpenter TC, Erdmann J, Franconi F, Giannetta E, Glezerman M, Hofmann SM, Junien C, Katai M, Kublickiene K, König IR, Majdic G, Malorni W, Mieth C, Miller VM, Reynolds RM, Shimokawa H, Tannenbaum C, D'Ursi AM, Regitz-Zagrosek V. (2017). Sex in basic research: Concepts in the cardiovascular field. *Cardiovasc Res* 113(7): 711–724.

185 *For many years, women had been taking drugs like the antihistamine*: To access a list of the prescription drugs that were withdrawn from the U.S. market between 1997 and 2000 because of evidence of greater health risks for women, see the following website: https://www.gao.gov/new.items/d01286r.pdf.

185 *We know that it takes genetic females much longer to clear*: Waxman DJ, Holloway MG. (2009). Sex differences in the expression of hepatic drug metabolizing enzymes. *Mol Pharmacol* 76(2): 215–228; Whitley HP, Lindsey W. (2009). Sex-based differences in drug activity. *Am Fam Physician* 80(11): 1254–1258.

185 *What this means practically for women is that they need to wait*: Beierle I, Meibohm B, Derendorf H. (1999). Gender differences in pharmacokinetics and pharmacodynamics. *Int J Clin Pharmacol Ther* 37(11): 529–547; Datz FL, Christian PE, Moore J. (1987). Gender-related differences in gastric emptying. *J Nucl Med* 28(7): 1204–1207; Islam MM, Iqbal U, Walther BA, Nguyen PA, Li YJ, Dubey NK, Poly TN, Masud JHB, Atique S, Syed-Abdul S. (2017). Gender-based personalized pharmacotherapy: A systematic review. *Arch Gynecol Obstet* 295(6): 1305–1317.

186 *Zelnorm (tegaserod) is one such drug*: Chey WD, Paré P, Viegas A, Ligozio G, Shetzline MA. (2008). Tegaserod for female patients suffering from IBS with mixed bowel habits or constipation: A randomized controlled trial. *Am J Gastroenterol* 103(5): 1217–1225; Tack J, Müller-Lissner S,

Bytzer P, Corinaldesi R, Chang L, Viegas A, Schnekenbuehl S, Dunger-Baldauf C, Rueegg P. (2005). A randomised controlled trial assessing the efficacy and safety of repeated tegaserod therapy in women with irritable bowel syndrome with constipation. *Gut* 54(12): 1707–1713; Wagstaff AJ, Frampton JE, Croom KF. (2003). Tegaserod: A review of its use in the management of irritable bowel syndrome with constipation in women. *Drugs* 63(11): 1101–1120.

186 *Similarly, at a recent meeting of the NIH Advisory Committee*: McCullough LD, Zeng Z, Blizzard KK, Debchoudhury I, Hurn PD. (2005). Ischemic nitric oxide and poly (ADP-ribose) polymerase-1 in cerebral ischemia: Male toxicity, female protection. *J Cereb Blood Flow Meta* 25(4): 502–512. To read more, see National Institutes of Health. Sex as biological variable: A step toward stronger science, better health. https://orwh.od.nih.gov/about/director/messages/sex-biological-variable.

188 *Her symptoms didn't really sound like the symptoms of stress incontinence*: Schierlitz L, Dwyer PL, Rosamilia A, Murray C, Thomas E, De Souza A, Hiscock R. (2012). Three-year follow-up of tension-free vaginal tape compared with transobturator tape in women with stress urinary incontinence and intrinsic sphincter deficiency. *Obstet Gynecol* 119(2 Pt 1): 321–327; Kalejaiye O, Vij M, Drake MJ. (2015). Classification of stress urinary incontinence. *World J Urol* 33(9): 1215–1220.

188 *It turned out that the real "problem" was not incontinence, but female ejaculation*: For more information about female ejaculation, see the following book and article: Sharon Moalem. (2009). *How Sex Works: Why We Look, Smell, Taste, Feel and Act the Way We Do*. New York: HarperCollins; Wimpissinger F, Stifter K, Grin W, Stackl W. (2007). The female prostate revisited: Perineal ultrasound and biochemical studies of female ejaculate. *J Sex Med* 4(5): 1388–1393.

189 *More than fifteen hundred years ago, however, both Aristotle*: Korda JB, Goldstein SW, Sommer F. (2010). The history of female ejaculation. *J Sex Med* 7(5): 1965–1675; Moalem S, Reidenberg JS. (2009). Does female ejaculation serve an antimicrobial purpose? *Med Hypotheses* 73(6): 1069–1071.

189 *In the seventeenth century, the Dutch anatomist and physician*: Biancardi MF, Dos Santos FCA, de Carvalho HF, Sanches BDA, Taboga SR. (2017). Female prostate: Historical, developmental, and morphological perspectives. *Cell Biol Int* 41(11): 1174–1183; Korda JB, Goldstein SW, Sommer F. (2010). The history of female ejaculation. *J Sex Med* 7(5): 1965–1975.

189 *The Englishman William Smellie, who was practicing*: Heath D. (1984). An

investigation into the origins of a copious vaginal discharge during intercourse: "Enough to wet the bed": That "is not urine." *J Sex Res* 20(2): 194–210.

190 *When you look up Skene in a clinical anatomy textbook*: John T. Hansen (2018). *Netter's Clinical Anatomy*. New York: Elsevier; Wimpissinger F, Tscherney R, Stackl W. (2009). Magnetic resonance imaging of female prostate pathology. *J Sex Med* 6(6): 1704–1711.

190 *In 2001, the Federative International Committee*: Sharon Moalem. (2009). *How Sex Works: Why We Look, Smell, Taste, Feel and Act the Way We Do*. New York: HarperCollins.

190 *We now also know that fluid released by some women*: Dietrich W, Susani M, Stifter L, Haitel A. (2011). The human female prostate-immunohistochemical study with prostate-specific antigen, prostate-specific alkaline phosphatase, and androgen receptor and 3-D remodeling. *J Sex Med* 8(10): 2816–2821.

192 *The urologist I referred her to specialized in urologic oncology*: As mentioned in the text, there have been extremely rare case reports of females developing Skene's duct adenocarcinoma (female prostate cancer), some with a concomitant elevation of serum prostate-specific antigen (PSA) level. To read more about this topic, see the following articles: Dodson MK, Cliby WA, Keeney GL, Peterson MF, Podratz KC. (1994). Skene's gland adenocarcinoma with increased serum level of prostate-specific antigen. *Gynecol Oncol* 55(2): 304–307; Korytko TP, Lowe GJ, Jimenez RE, Pohar KS, Martin DD. (2012). Prostate-specific antigen response after definitive radiotherapy for Skene's gland adenocarcinoma resembling prostate adenocarcinoma. *Urol Oncol* 30(5): 602–606; Pongtippan A, Malpica A, Levenback C, Deavers MT, Silva EG. (2004). Skene's gland adenocarcinoma resembling prostatic adenocarcinoma. *Int J Gynecol Pathol* 23(1): 71–74; Tsutsumi S, Kawahara T, Hattori Y, Mochizuki T, Teranishi JI, Makiyama K, Miyoshi Y, Otani M, Uemura H. (2018). Skene duct adenocarcinoma in a patient with an elevated serum prostate-specific antigen level: A case report. *J Med Case Rep* 12(1): 32; Zaviacic M, Ablin RJ. The female prostate and prostate-specific antigen. (2000). Immunohistochemical localization, implications of this prostate marker in women and reasons for using the term "prostate" in the human female. *Histol Histopathol* 15(1): 131–142.

194 *Specifically, my research focused on what was at the time*: Moalem S, Weinberg ED, Percy ME. (2004). Hemochromatosis and the enigma of misplaced iron: Implications for infectious disease and survival. *Biometals*

17(2): 135–139; Galaris D, Pantopoulos K. (2008). Oxidative stress and iron homeostasis: Mechanistic and health aspects. *Crit Rev Clin Lab Sci* 45(1): 1–23; Kander MC, Cui Y, Liu Z. (2017). Gender difference in oxidative stress: A new look at the mechanisms for cardiovascular diseases. *J Cell Mol Med* 21(5): 1024–1032; Pilo F, Angelucci E. (2018). A storm in the niche: Iron, oxidative stress and haemopoiesis. *Blood Rev* 32(1): 29–35.

194 *The gene associated with hemochromatosis*: Feder JN, Gnirke A, Thomas W, Tsuchihashi Z, Ruddy DA, Basava A, Dormishian F, Domingo R Jr, Ellis MC, Fullan A, Hinton LM, Jones NL, Kimmel BE, Kronmal GS, Lauer P, Lee VK, Loeb DB, Mapa FA, McClelland E, Meyer NC, Mintier GA, Moeller N, Moore T, Morikang E, Prass CE, Quintana L, Starnes SM, Schatzman RC, Brunke KJ, Drayna DT, Risch NJ, Bacon BR, Wolff RK. (1996). A novel MHC class I–like gene is mutated in patients with hereditary haemochromatosis. *Nat Genet* 13(4): 399–408.

195 *This is because most genetic females lose iron*: For more on the biological relationship between iron and human health, see the following book: Sharon Moalem with Jonathan M. Prince. (2007). *Survival of the Sickest: A Medical Maverick Discovers Why We Need Disease*. New York: William Morrow.

196 *To this day, the treatment for hemochromatosis*: Although there are new treatments currently being tested, besides dietary modifications, phlebotomy or bloodletting is still the most common treatment used today for hemochromatosis. For more on this topic, see the following book and articles: Robert Root-Bernstein, Michele Root-Bernstein. (1997). *Honey, Mud, Maggots, and Other Medical Marvels: The Science Behind Folk Remedies and Old Wives' Tales*. Boston: Houghton Mifflin; Kowdley KV, Brown KE, Ahn J, Sundaram V. (2019). ACG Clinical Guideline: Hereditary Hemochromatosis. *AM J Gastroenterol* 114(8): 1202–1218. For a paper covering the history of bloodletting, see Seigworth GR. (1980). Bloodletting over the centuries. *NY State J Med* 80(13): 2022–2028.

199 *The reason Venerina's ventricles are of equal thickness*: Mazzotti G, Falconi M, Teti G, Zago M, Lanari M, Manzoli FA. (2010). The diagnosis of the cause of the death of Venerina. *J Anat* 216(2): 271–274.

200 *Studies have found much variability in a physician's ability*: Wernli KJ, Henrikson NB, Morrison CC, Nguyen M, Pocobelli G, Blasi PR. (2016). Screening for skin cancer in adults: Updated evidence report and systematic review for the US Preventive Services Task Force. *JAMA* 316(4): 436–447; Wernli KJ, Henrikson NB, Morrison CC, Nguyen M,

Pocobelli G, Whitlock EP. (2016). Screening for skin cancer in adults: An updated systematic evidence review for the U.S. Preventive Services Task Force. *USPSTF: Agency for Healthcare Research and Quality.* Available from http://www.ncbi.nlm.nih.gov/books/NBK379854/.

200 *Malignant melanoma is the least common type of skin cancer*: Geller AC, Miller DR, Annas GD, Demierre MF, Gilchrest BA, Koh HK. (2002). Melanoma incidence and mortality among US whites, 1969–1999. *JAMA* 288(14): 1719–1720; Rastrelli M, Tropea S, Rossi CR, Alaibac M. (2014). Melanoma: Epidemiology, risk factors, pathogenesis, diagnosis and classification. *In Vivo* 28(6): 1005–1011.

201 *This is why the location of melanoma is different*: Although it is still possible to develop melanoma in non-sun-exposed areas of the body (the oral cavity and nasal sinuses, for example), the general rationale behind the location of the majority of melanoma cases is increased exposure to UV. See the following articles for more information on this topic: Chevalier V, Barbe C, Le Clainche A, Arnoult G, Bernard P, Hibon E, Grange F. (2014). Comparison of anatomical locations of cutaneous melanoma in men and women: A population-based study in France. *Br J Dermatol* 171(3): 595–601; Lesage C, Barbe C, Le Clainche A, Lesage FX, Bernard P, Grange F. (2013). Sex-related location of head and neck melanoma strongly argues for a major role of sun exposure in cars and photoprotection by hair. *J Invest Dermatol* 133(5): 1205–1211.

201 *colorectal cancer is more common in males*: Chacko L, Macaron C, Burke CA. (2015). Colorectal cancer screening and prevention in women. *Dig Dis Sci* 60(3): 698–710; Li CH, Haider S, Shiah YJ, Thai K, Boutros PC. (2018). Sex differences in cancer driver genes and biomarkers. *Cancer Res* 78(19): 5527–5537; Radkiewicz C, Johansson ALV, Dickman PW, Lambe M, Edgren G. (2017). Sex differences in cancer risk and survival: A Swedish cohort study. *Eur J Cancer* 84: 130–140.

202 *We're starting to realize that traumatic brain injuries*: Resch JE, Rach A, Walton S, Broshek DK. (2017). Sport concussion and the female athlete. *Clin Sports Med* 36(4): 717–739; Covassin T, Moran R, Elbin RJ. (2016). Sex differences in reported concussion injury rates and time loss from participation: An update of the National Collegiate Athletic Association Injury Surveillance Program from 2004–2005 through 2008–2009. *J Athl Train* 51(3): 189–194.

204 *Studies looking at sports with similar rules for both women and men*: Dick RW. (2009). Is there a gender difference in concussion incidence and

outcomes? *Br J Sports Med* Suppl 1: i46–50; Yuan W, Dudley J, Barber Foss KD, Ellis JD, Thomas S, Galloway RT, DiCesare CA, Leach JL, Adams J, Maloney T, Gadd B, Smith D, Epstein JN, Grooms DR, Logan K, Howell DR, Altaye M, Myer GD. (2018). Mild jugular compression collar ameliorated changes in brain activation of working memory after one soccer season in female high school athletes. *J Neurotrauma* 35(11): 1248–1259.

204 *In addition, the physical proportions of the neck and head*: Resch JE, Rach A, Walton S, Broshek DK. (2017). Sport concussion and the female athlete. *Clin Sports Med* 36(4): 717–739; Tierney RT, Sitler MR, Swanik CB, Swanik KA, Higgins M, Torg J. (2005). Gender differences in head-neck segment dynamic stabilization during head acceleration. *Med Sci Sports Exerc* 37(2): 272–279.

206 *For a woman, any hemoglobin level below 12 grams per deciliter is associated with anemia*: J. Larry Jameson, Anthony S. Fauci, Dennis L. Kasper, Stephen L. Hauser, Dan L. Longo, Joseph Loscalzo. (2018). *Harrison's Principles of Internal Medicine*. Vols. 1 and 2. 20th ed. New York: McGraw-Hill Education.

208 *Even though the studies have been small so far*: Harmon KG, Drezner JA, Gammons M, Guskiewicz KM, Halstead M, Herring SA, Kutcher JS, Pana A, Putukian M, Roberts WO. (2013). American Medical Society for Sports Medicine position statement: Concussion in sport. 47(1): 15–26.

209 *An illustrated example of this can be seen from the results*: Sollmann N, Echlin PS, Schultz V, Viher PV, Lyall AE, Tripodis Y, Kaufmann D, Hartl E, Kinzel P, Forwell LA, Johnson AM, Skopelja EN, Lepage C, Bouix S, Pasternak O, Lin AP, Shenton ME, Koerte IK. (2017). Sex differences in white matter alterations following repetitive subconcussive head impacts in collegiate ice hockey players. *Neuroimage Clin* 17: 642–649.

210 *In the field of cardiovascular disease, it is established*: Chauhan A, Moser H, McCullough LD. (2017). Sex differences in ischaemic stroke: Potential cellular mechanisms. *Clin Sci* 131(7): 533–552; King A. (2011). The heart of a woman: Addressing the gender gap in cardiovascular disease. *Nat Rev Cardiol* 8(11): 239–240; Angela H.E.M. Maas, C. Noel Bairey Merz, eds. (2017). *Manual of Gynecardiology: Female-Specific Cardiology*. New York: Springer; Regitz-Zagrosek V, Kararigas G. (2017). Mechanistic pathways of sex differences in cardiovascular disease. *Physiol Rev* 97(1): 1–37.

213 *More than 90 percent of people diagnosed with takotsubo cardiomyopathy*:

Takotsubo was first described by Dr. Hiraku Sato in 1990 in a publication written in Japanese. See the following articles for more information on this topic: Berry D. (2013). Dr. Hikaru Sato and Takotsubo cardiomyopathy or broken heart syndrome. *Eur Heart J* 34(23): 1695; Kurisu S, Kihara Y. (2012). Tako-tsubo cardiomyopathy: Clinical presentation and underlying mechanism. *J Cardiol* 60(6): 429–37; Tofield A. (2016). Hikaru Sato and Takotsubo cardiomyopathy. *Eur Heart J* 37(37): 2812.

214 *Around one hundred thousand people are currently waiting*: For the most up-to-date list of the number of patients waiting for an organ transplant in the United States, see the following website: https://unos.org/data /transplant-trends/waiting-list-candidates-by-organ-type/.

214 *Overall, male kidneys contain more nephrons*: Tsuboi N, Kanzaki G, Koike K, Kawamura T, Ogura M, Yokoo T. (2014). Clinicopathological assessment of the nephron number. *Clin Kidney J* 7(2): 107–114.

215 *The majority of the people in need of a new kidney*: Lai Q, Giovanardi F, Melandro F, Larghi Laureiro Z, Merli M, Lattanzi B, Hassan R, Rossi M, Mennini G. (2018). Donor-to-recipient gender match in liver transplantation: A systematic review and meta-analysis. *World J Gastroenterol* 24(20): 2203–2210; Puoti F, Ricci A, Nanni-Costa A, Ricciardi W, Malorni W, Ortona E. (2016). Organ transplantation and gender differences: A paradigmatic example of intertwining between biological and sociocultural determinants. *Biol Sex Differ* 7: 35.

215 *Several clinical trials have found that receiving*: Martinez-Selles M, Almenar L, Paniagua-Martin MJ, Segovia J, Delgado JF, Arizón JM, Ayesta A, Lage E, Brossa V, Manito N, Pérez-Villa F, Diaz-Molina B, Rábago G, Blasco-Peiró T, De La Fuente Galán L, Pascual-Figal D, Gonzalez-Vilchez F; Spanish Registry of Heart Transplantation. (2015). Donor/recipient sex mismatch and survival after heart transplantation: Only an issue in male recipients? An analysis of the Spanish Heart Transplantation Registry. *Transpl Int* 28(3): 305–313; Zhou JY, Cheng J, Huang HF, Shen Y, Jiang Y, Chen JH. (2013). The effect of donor-recipient gender mismatch on short- and long-term graft survival in kidney transplantation: A systematic review and meta-analysis. *Clin Transplant* 27(5): 764–771.

216 *Some of these differences in outcome*: Foroutan F, Alba AC, Guyatt G, Duero Posada J, Ng Fat Hing N, Arseneau E, Meade M, Hanna S, Badiwala M, Ross H. (2018). Predictors of 1-year mortality in heart transplant recipients: A systematic review and meta-analysis. *Heart* 104(2): 151–160; Puoti F, Ricci A, Nanni-Costa A, Ricciardi W, Malorni W, Ortona E. (2016). Organ transplantation and gender differences: A

paradigmatic example of intertwining between biological and sociocultural determinants. *Biol Sex Differ* 7: 35.

Conclusion

222 *The bacteria* Salmonella typhi *can cause a dreadful infection*: Yang YA, Chong A, Song J. (2018). Why is eradicating typhoid fever so challenging: implications for vaccine and therapeutic design. *Vaccines (Basel)* 6(3). See the following World Health Organization's site for more information as well: https://www.who.int/ith/vaccines/typhoidfever/en/.

222 *Based on the differences between our physical reaction to the vaccine*: Fischinger S, Boudreau CM, Butler AL, Streeck H, Alter G. (2019). Sex differences in vaccine-induced humoral immunity. *Semin Immunopathol* 41(2): 239–249; Flanagan KL, Fink AL, Plebanski M, Klein SL. (2017). Sex and gender differences in the outcomes of vaccination over the life course. *Annu Rev Cell Dev Biol* 33: 577–599; Giefing-Kröll C, Berger P, Lepperdinger G, Grubeck-Loebenstein B. (2015). How sex and age affect immune responses, susceptibility to infections, and response to vaccination. *Aging Cell* 14(3): 309–321; Schurz H, Salie M, Tromp G, Hoal EG, Kinnear CJ, Möller M. (2019). The X chromosome and sex-specific effects in infectious disease susceptibility. *Hum Genomics* 13(1): 2.

223 *Recent research has discovered that genetic females and males use*: Gershoni M, Pietrokovski S. (2017). The landscape of sex-differential transcriptome and its consequent selection in human adults. *BMC Biol* 15(1): 7.

224 *Let's review the numbers: Genetic males begin life*: The sex-ratio-at-birth data is available for almost all countries worldwide and is kept by the UN; see the following website: http://data.un.org/Data.aspx?d=PopDiv&f=variableID%3A52.

224 *Around the time people reach the age of 40, the number of females and males*: For more research articles, see Dulken B, Brunet A. (2015). Stem cell aging and sex: Are we missing something? *Cell Stem Cell* 16(6): 588–590; Marais GAB, Gaillard JM, Vieira C, Plotton I, Sanlaville D, Gueyffier F, Lemaitre JF. (2018). Sex gap in aging and longevity: Can sex chromosomes play a role? *Biol Sex Differ* 9(1); Ostan R, Monti D, Gueresi P, Bussolotto M, Franceschi C, Baggio G. (2016). Gender, aging and longevity in humans: An update of an intriguing/neglected scenario paving the way to a gender-specific medicine. *Clin Sci (Lond)* 130(19): 1711–1725. For more on the demographic landscape between females and males, see https://unstats.un.org/unsd/gender/downloads/WorldsWomen2015_chapter1_t.pdf.

225 *Of the fifteen leading causes of death in the United States*: Since the adult sex ratio for older individuals is not 1:1, comparisons between the sexes such as this are always matched for age and sex. See the following study for more details: Austad SN, Fischer KE. (2016). Sex differences in lifespan. *Cell Metab* 23(6): 1022–1033.

225 *Tanaka's life history is an illustration of the female survival advantage*: If you'd like to see a few photographs and read more about Tanaka's amazing life, see the following websites: http://www.guinnessworldrecords .com/news/2019/3/worlds-oldest-person-confirmed-as-116-year-old -kane-tanaka-from-japan; https://www.bbc.com/news/video_and_audio /headlines/47508517/oldest-living-person-kane-tanaka-celebrates -getting-the-guinness-world-record.

226 *Recent research studying adult sex ratios in 344 species*: Pipoly I, Bokony V, Kirkpatrick M, Donald PF, Szekely T, Liker A. (2015). The genetic sex-determination system predicts adult sex ratios in tetrapods. *Nature* 527(7576): 91–94.

ACKNOWLEDGMENTS

I am incredibly thankful to all my research participants and patients who gave so generously of themselves (their generosity included donations of time, blood, tissue, genetic material, and facial scans), which helped make many of the studies described in this book and my research studying the differences between the genetic sexes over the years possible. As well, my gratitude to all the rare disease funding and government agencies, who helped support my research work. Of my many research colleagues, postdoctoral students, and graduate students, I would like to particularly thank Professor Jason T. C. Tzen (曾志正) of the National Chung Hsing University (NCHU). My discussions and correspondences with Jason regarding the related and divergent biological coping strategies of plants and animals to extreme environmental stressors across many species, which have taken place over many years, were fruitful for me as the theoretical premise of this book began to take shape. As our conversations always reminded me, there's still so much that we have yet to discover about *Camellia sinensis sinensis*, no matter where in the world it grows.

A special mention to the incredibly thoughtful hospitality that the Centro Internacional de la Papa (CIP) has always shown me. Thank you to Maria Elena Lanatta at CIP, whose stellar organizational know-how provided guidance and structure to my visits and work with all the staff. Because so many of the world's important crops, such as potatoes, originate in the Andes, the work of CIP to protect Andean biodiversity and thus provide food security for us all has never been more essential. If there's anything we can learn from history, it's that we can never be too prepared to deal with the next time a pathogen ravages our crops and famine returns.

In that regard I continue to be impressed by the work of the Guardians of the Potato. Unlike many other crops, potatoes are propagated using seed-potato, and so the storing and protection of the thousands of varieties of potatoes by the Guardians is an added fundamental insurance policy for all of humanity. As they always have, the Guardians provided me with impeccable hospitality, and the bounty of the offerings they shared with me during my visits was extremely moving. And I wouldn't have been able to do any of this crucial research and work without the Asociación ANDES, who provided essential and professional coordinating and scheduling. Alejandro Argumedo did so much more than just guide me seamlessly through the picturesque Sacred Valley and the astonishing Peruvian Potato Park. Our conversations, which took place over long drives through some pretty scenic and formidable backdrops, provided much for me to reflect upon further as I contemplated the relationship between adverse growing conditions and biological resiliency and questioned what polyploid potatoes could teach us about human genetics.

And to Chef Yoshihiro Murata of Kikunoi, who is unrivaled in his masterful command of Japanese cuisine as a whole. In particular, Chef Murata's detailed understanding of a little-known ancient fishing practice sparked my thinking about some of the differences in chromosomal sex determination between avian and mammalian species and their implications in a whole new light. Our conversations about historical and contemporary food growing practices in Japan, and the world over, were extremely enriching. A special thanks to all the farmers and producers and small-scale artisans who shared their time and expertise. Their inspiring devotion to tirelessly producing all the specialized ingredients needed for *washoku* continues to support and enable the existence of Japanese cuisine.

Of course, I am thankful to all my current and past staff, both research and operational, who provided the necessary support for me to be able to find the time to write. For her utmost efficiency and dedication, a special thanks to my executive assistant, Claire Matthews. There was a lot to juggle across numerous time zones, and with a few super typhoons and a major earthquake thrown in the mix in the last few months as I traveled and wrote and researched simultaneously, I'm very grateful she was always there to keep me on track and on time (or mostly so) for every flight, meeting, and deadline.

Without my editor at Farrar, Straus and Giroux, Colin Dickerman, who championed this project from the very beginning, this book would not have happened. Colin, your patience as I worked to clear out my schedule to carve out the necessary time required to write was legendary. I am especially appreciative of your rare combined skillset of exemplary line editing coupled with your high-level thematic editorial comments, which were both enriching in shaping the manuscript as it evolved over time. I continue to be impressed by everyone I have had the opportunity to meet and work with at FSG, and in that regard, I would like to express particular gratitude to the rest of the FSG core team that was intimately involved in this book, including: editorial assistant Ian Van Wye, production editor Janine Barlow, jacket designer Rodrigo Corral, interior designer Richard Oriolo, director of foreign rights Devon Mazzone, and finally my publicist, Lottchen Shivers.

As well, to my Penguin UK editor, Laura Stickney, whose enthusiasm about this project was palpable and contagious right from our very first introduction, and whose insightful feedback and editorial comments on the manuscript were vital. I am also immensely grateful to the rest of the talented UK core team, which included editorial assistant Holly Hunter, marketing manager Julie Woon, and my UK publicist, Kate Smith.

A heartfelt thanks to my very talented editor, Amanda Moon. Amanda, I am so grateful for your astute advice and careful eye.

Your thoughtful, persistent, and probing questions and all your suggestions throughout were very valuable. Thank you for challenging me to bring that added layer of clarity and connection for the reader.

I am also grateful to the many early readers of the manuscript who provided insightful feedback and comments. Especially to Professor Han Brunner, whose important scientific work was formative for me as I thought and wrote about the relationship between human genetics, brain development, and the consequences of having only one X chromosome. I am also grateful to Dr. Brunner for taking the time to go through the manuscript so thoughtfully, with particular attention to the areas that fall within his domain of expertise. Thank you to Professor William J. Sullivan, who spared absolutely nothing in terms of the time and intellectual energy he devoted to providing me with his truly helpful early thoughts, comments, and feedback on the manuscript while it was still in its preliminary stages. Bill, you are a passionate and talented science communicator, and a very generous human being.

As always to my agent and good friend Richard Abate of 3 Arts. Richard, I am continuously grateful for your wise counsel and creative acumen, and for your sense of humor, which has helped to keep me grounded and focused throughout this and all of our other projects together. To the indispensable Rachel Kim of 3 Arts, who tirelessly ensured that we stayed on schedule, organized, and reasonably punctual when it came to meeting our many deadlines.

To all my friends, family, and famiglia, for your unfailing love and support over the years. And finally, it goes without saying that this book was inspired by, and dedicated to, my better half.